T0260337

Modern Data Visualization with R

Modern Data Visualization with R describes the many ways that raw and summary data can be turned into visualizations that convey meaningful insights. It starts with basic graphs such as bar charts, scatter plots, and line charts, but progresses to less well-known visualizations such as tree maps, alluvial plots, radar charts, mosaic plots, effects plots, correlation plots, biplots, and the mapping of geographic data. Both static and interactive graphics are described, and the use of color, shape, shading, grouping, annotation, and animations are covered in detail. The book moves from a default look and feel for graphs to graphs with customized colors, fonts, legends, annotations, and organizational themes.

Features

- Contains a wide breadth of graph types including newer and less well-known approaches
- Connects each graph type to the characteristics of the data and the goals of the analysis
- Moves the reader from simple graphs, describing one variable, to building visualizations that describe complex relationships among many variables
- Provides newer approaches to creating interactive web graphics via JavaScript libraries
- Details how to customize each graph type to meet users' needs and those of their audiences
- Gives methods for creating visualizations that are publication-ready (in color or black and white) and the web
- Suggests best practices
- Offers examples from a wide variety of fields

The text is written for those new to data analysis as well as seasoned data scientists. It can be used for both teaching and research, and will particularly appeal to anyone who needs to describe data visually and wants to find and emulate the most appropriate method quickly. The reader should have some basic coding experience, but expertise in R is not required. Some of the later chapters (e.g., visualizing statistical models) assume exposure to statistical inference at the level of analysis of variance and regression.

Robert Kabacoff is a data scientist with more than 30 years of experience in multivariate statistical methods, data visualization, predictive analytics, and psychometrics. A widely recognized expert in statistical programming, he is the author of *R in Action: Data Analysis and Graphics with R (3rd ed.)*, and the popular *Quick-R* (www.statmethods.net) website. Dr. Kabacoff is also the co-author of *Evaluating Research Articles from Start to Finish (3rd ed.)*, a textbook that uses a case study approach to help students learn to read and evaluate empirical research. He earned his BA in psychology from the University of Connecticut and his PhD in clinical psychology from the University of Missouri-St. Louis. Following a postdoctoral fellowship in family research at Brown University, he joined the faculty at the Center for Psychological Studies at Nova Southeastern University, achieving the position of full professor in 1997. For 19 years, Dr. Kabacoff held the position of Vice President of Research for a global organizational development firm, providing research and consultation to academic, government, corporate, and humanitarian institutions in North America, Western Europe, Africa, and Asia. He is currently a professor of the practice in quantitative analysis at the Quantitative Analysis Center at Wesleyan University, teaching courses in exploratory data analysis, machine learning, and statistical software development.

Chapman & Hall/CRC
The R Series

Series Editors

John M. Chambers, Department of Statistics, Stanford University, California, USA

Torsten Hothorn, Division of Biostatistics, University of Zurich, Switzerland

Duncan Temple Lang, Department of Statistics, University of California, Davis, USA

Hadley Wickham, RStudio, Boston, Massachusetts, USA

Recently Published Titles

For more information about this series, please visit: https://www.crcpress.com/Chapman--HallCRC-The-R-Series/book-series/CRCTHERSER

Modern Data Visualization with R

Robert Kabacoff

CRC Press
Taylor & Francis Group
Boca Raton London New York

CRC Press is an imprint of the
Taylor & Francis Group, an **informa** business

A CHAPMAN & HALL BOOK

First edition published 2024
by CRC Press
2385 Executive Center Drive, Suite 320, Boca Raton, FL 33431, U.S.A.

and by CRC Press
4 Park Square, Milton Park, Abingdon, Oxon, OX14 4RN

CRC Press is an imprint of Taylor & Francis Group, LLC

© 2024 Robert Kabacoff

ISBN: 978-1-0322-8949-6 (hbk)
ISBN: 978-1-0322-8760-7 (pbk)
ISBN: 978-1-0032-9927-1 (ebk)

DOI: 10.1201/9781003299271

Typeset in LM Roman
by KnowledgeWorks Global Ltd.

Publisher's note: This book has been prepared from camera-ready copy provided by the authors.

Contents

Preface

*There is magic in graphs. The profile of a curve reveals in a flash
a whole situation–the life history of an epidemic, a panic, or an era
of prosperity. The curve informs the mind, awakens the imagination,
convinces.*

–Henry D. Hubbard

Above all else, show the data.

–Edward Tufte

*We cannot just look at a country by looking at charts, graphs, and
modeling the economy. Behind the numbers there are people.*

–Christine Lagarde

R is an amazing platform for data analysis, capable of creating almost any type
of graph. This book helps you create the most popular visualizations–from
quick and dirty plots to publication-ready graphs. The text relies heavily on
the **ggplot2** (https://ggplot2.tidyverse.org/) package for graphics, but other
approaches are covered as well.

0.1 Why This Book?

There are many books on data visualization using R. So why another one? I
am trying to achieve five goals with this book.

(1) **Help identify the most appropriate graph for a given situa-
    tion**. With the plethora of types of graphs available, some guidance
    is required when choosing a graph for a given problem. I've tried to
    provide that guidance here.

(2) **Allow easy access to these graphs**. The graphs in this book
    are presented in **cookbook** fashion. Basic graphs are demonstrated
    first, followed by more attractive and customized versions.

(3) **Expand the breadth of visualizations available**. There are
    many more types of graphs than we typically see in reports and
    blogs. They can be helpful, intuitive, and compelling. I've tried to
    include many of them here.

(4) **Help you customize any graph to meet your needs**. Basic graphs are easy, but highly customized graphs can take some work. This book provides the necessary details for modifying axes, shapes, colors, fonts, annotations, formats, and more. You can make your graph look *exactly* as you wish.

(5) **Offer suggestions for best practices**. There is an ethical obligation to convey information clearly, and with as little distortion or obfuscation as possible. I hope this book helps support that goal.

0.2 Acknowledgments

I want to acknowledge the people at CRC Press and KGL who helped to make this book possible.

There are two other people I would like to thank. The first person is Manolis Kaparakis, Director of the Quantitative Analysis Center at Wesleyan University and ostensibly my boss. He has always strived to empower me and help me feel valued and appreciated. He is simply the best boss I've ever had. We should all be so lucky.

The second person is really the first person in all things. It was my idea to write this book. It was my wife Carol Lynn's idea to finish the book. Her love and support knows no bounds, and this book is a statistician's version of PDA. How did I get so lucky?

0.3 Supporting Website

An online version of this book is available at http://rkabacoff.github.io/datavis.

Supplementary materials (including all the code and datasets used in this book) are available on the support website, http://github.com/rkabacoff/datavis_support.

1

Introduction

If you are reading this book, you probably already appreciate the importance of visualizing data. It is an essential component of any data analysis. While we generally accept the old adage that *"a picture is worth a thousand words"*, it is worthwhile taking a moment to consider why.

Humans are remarkably capable of discerning patterns in visual data. This allows us to discover relationships, identify unusual or mistaken values, determine trends and differences, and with some effort, understand the relationships among several variables at once.

Additionally, data visualizations tend to have a greater cognitive and *emotional* impact than either text descriptions or tables of numbers. This makes them a key ingredient in both storytelling and the crafting of persuasive arguments.

Because graphs are so compelling, researchers and data scientists have an ethical obligation to create visualizations that fairly and accurately reflect the information contained in the data. The goal of this book is to provide you with the tools to both select and create graphs that present data as clearly, understandably, and accurately (honestly) as possible.

The R platform (R Core Team, 2023) provides one of the most comprehensive set of tools for accomplishing these goals. The software is open source, freely available, runs on almost any platform, is highly customizable, and is supported by a massive worldwide user base. The tools described in this book should allow you to create almost any type of data visualization desired.

Currently, the most popular approach to creating graphs in R uses **ggplot2** package (Wickham et al., 2023). Based on a *Grammar of Graphics* (Wilkinson & Wills, 2005), the ggplot2 package provides a coherent and extensible system for data visualization and is the central approach used in this book. Since its release, a number of additional packages have been developed to enhance and expand the types of graphs that can easily be created with ggplot2. Many of these are explored in later chapters.

1.1   How to Use This Book

I hope that this book will provide you with a comprehensive overview of data visualization. However, you don't need to read this book from start to finish in

DOI: 10.1201/9781003299271-1

order to start building effective graphs. Feel free to jump to the section that you need, and then explore others that you find interesting.

Graphs are organized by

- the number of variables to be plotted
- the type of variables to be plotted
- the purpose of the visualization

Chapter	Description
Ch 2	Provides a quick overview of how to get your data into R and how to prepare it for analysis.
Ch 3	Provides an overview of the **ggplot2** package.
Ch 4	Describes graphs for visualizing the distribution of a single categorical (e.g., race) or quantitative (e.g., income) variable.
Ch 5	Describes graphs that display the relationship between two variables.
Ch 6	Describes graphs that display the relationships among three or more variables. It is helpful to read Chapters 4 and 5 before this chapter.
Ch 7	Provides a brief introduction to displaying data geographically.
Ch 8	Describes graphs that display change over time.
Ch 9	Describes graphs that can help you interpret the results of statistical models.
Ch 10	Covers graphs that do not fit neatly elsewhere (every book needs a miscellaneous chapter).
Ch 11	Describes how to customize the look and feel of your graphs. If you are going to share your graphs with others, be sure to check it out.
Ch 12	Covers how to save your graphs. Different formats are optimized for different purposes
Ch 13	Provides an introduction to interactive graphics.
Ch 14	Gives advice on creating effective graphs and where to go to learn more. It's worth a look.
The Appendices	Describe each of the datasets used in this book, and provide a short blurb about the author and the Wesleyan QAC.

There is **no one right graph** for displaying data. Check out the examples, and see which type best fits your needs.

1.2 Pre-requisites

It's assumed that you have some experience with the R language and that you have already installed R and RStudio. If not, here are two excellent resources for getting started:

- **A (very) short introduction to R** by Paul Torfs and Claudia Brauer (https://cran.r-project.org/doc/contrib/Torfs+Brauer-Short-R-Intro. pdf). This is a great introductory article that will get you up and running quickly.

- **An Introduction to R** by Alex Douglas, Deon Roos, Francesca Mancini, Anna Couto, and David Lussea (https://intro2r.com). This is a comprehensive e-book on R. Chapters 1–3 provide a solid introduction.

Either of these resources will help you familiarize yourself with R quickly.

1.3 Setup

In order to create the graphs in this book, you'll need to install a number of optional R packages. Most of these packages are hosted on the **Comprehensive R Archive Network** (CRAN). To install **all** of these CRAN packages, run the following code in the RStudio console window.

```
CRAN_pkgs <- c("ggplot2", "dplyr", "tidyr", "mosaicData",
               "carData", "VIM", "scales", "treemapify",
               "gapminder","sf", "tidygeocoder", "mapview",
               "ggmap", "osmdata", "choroplethr",
               "choroplethrMaps", "lubridate", "CGPfunctions",
               "ggcorrplot", "visreg", "gcookbook", "forcats",
               "survival", "survminer", "car", "rgl",
               "ggalluvial", "ggridges", "GGally", "superheat",
               "waterfalls", "factoextra","networkD3",
               "ggthemes", "patchwork", "hrbrthemes", "ggpol",
               "quantmod", "gghighlight", "leaflet", "ggiraph",
               "rbokeh", "ggalt")
install.packages(CRAN_pkgs)
```

Alternatively, you can install a given package the first time it is needed. For example, if you execute
```
library(gapminder)
```
and get the message

```
Error in library(gapminder) : there is no package called
'gapminder'
```
you know that the package has never been installed. Simply execute
```
install.packages("gapminder")
```
once and
```
library(gapminder)
```
will work from that point on.

A few specialized packages used later in the book are only hosted on **GitHub**. You can install them using the `install_github` function in the **remotes** package. First install the **remotes** package from CRAN.

```
install.packages("remotes")
```

Then run the following code to install the remaining packages:

```
github_pkgs <- c("rkabacoff/ggpie", "hrbrmstr/waffle",
                 "ricardo-bion/ggradar", "ramnathv/rCharts",
                 "Mikata-Project/ggthemr")
remotes::install_github(github_pkgs, dependencies = TRUE)
```

Although it may seem like a lot, these packages should install fairly quickly. And again, you can install them individually as needed.

At this point, you should be ready to go. Let's get started!

2

Data Preparation

Before you can visualize your data, you have to get it into R. This involves importing the data from an external source and massaging it into a useful format. It would be great if data came in a clean rectangular format, without errors, or missing values. It would also be great if ice cream grew on trees. A significant part of data analysis is preparing the data for analysis.

2.1 Importing Data

R can import data from almost any source, including text files, Excel spreadsheets, statistical packages, and database management systems (DBMS). We'll illustrate these techniques using the `Salaries` dataset, containing the 9-month academic salaries of college professors at a single institution in 2008–2009. The dataset is described in Appendix A.1.

2.1.1 Text Files

The **readr** package provides functions for importing delimited text files into R data frames.

```
library(readr)

# import data from a comma delimited file
Salaries <- read_csv("salaries.csv")

# import data from a tab delimited file
Salaries <- read_tsv("salaries.txt")
```

These function assume that the first line of data contains the variable names, values are separated by commas or tabs respectively, and that missing data are represented by blanks. For example, the first few lines of the comma delimited file looks like this.

```
"rank","discipline","yrs.since.phd","yrs.service","sex","salary"
```

DOI: 10.1201/9781003299271-2   5

```
"Prof","B",19,18,"Male",139750
"Prof","B",20,16,"Male",173200
"AsstProf","B",4,3,"Male",79750
"Prof","B",45,39,"Male",115000
"Prof","B",40,41,"Male",141500
"AssocProf","B",6,6,"Male",97000
```

Options allow you to alter these assumptions. See the `?read_delim` for more details.

2.1.2   Excel Spreadsheets

The **readxl** package can import data from Excel workbooks. Both xls and xlsx formats are supported.

```
library(readxl)

# import data from an Excel workbook
Salaries <- read_excel("salaries.xlsx", sheet=1)
```

Since workbooks can have more than one worksheet, you can specify the one you want with the `sheet` option. The default is `sheet=1`.

2.1.3   Statistical Packages

The **haven** package provides functions for importing data from a variety of statistical packages.

```
library(haven)

# import data from Stata
Salaries <- read_dta("salaries.dta")

# import data from SPSS
Salaries <- read_sav("salaries.sav")

# import data from SAS
Salaries <- read_sas("salaries.sas7bdat")
```

Note: you do not need to have these statistical packages installed in order to import their data files.

2.1.4   Databases

Importing data from a database requires additional steps and is beyond the scope of this book. Depending on the database containing the data, the

TABLE 2.1

Package	Function	Use
dplyr	select	Select variables/columns
dplyr	filter	Select observations/rows
dplyr	mutate	Transform or recode variables
dplyr	summarize	Summarize data
dplyr	group_by	Identify subgroups for further processing
tidyr	gather	Convert wide format dataset to long format
tidyr	spread	Convert long format dataset to wide format

following packages can help: **RODBC**, **RMySQL**, **ROracle**, **RPostgreSQL**, **RSQLite**, and **RMongo**. In the newest versions of RStudio, you can use the Connections pane to quickly access the data stored in DBMS.

2.2 Cleaning Data

The processes of cleaning your data can be the most time-consuming part of any data analysis. The most important steps are considered below. While there are many approaches, those using the **dplyr** and **tidyr** packages are some of the quickest and easiest to learn.

Examples in this section will use the Star Wars dataset from the **dplyr** package. The dataset provides descriptions of 87 characters from the Star Wars universe on 13 variables. (I actually prefer Star Trek, but we work with what we have.) The dataset is described in Appendix A.2.

2.2.1 Selecting Variables

The select function allows you to limit your dataset to specified variables (columns).

```
library(dplyr)

# keep the variables name, height, and gender
newdata <- select(starwars, name, height, gender)

# keep the variables name and all variables
# between mass and species inclusive
newdata <- select(starwars, name, mass:species)

# keep all variables except birth_year and gender
newdata <- select(starwars, -birth_year, -gender)
```

2.2.2   Selecting Observations

The `filter` function allows you to limit your dataset to observations (rows) meeting a specific criteria. Multiple criteria can be combined with the & (AND) and | (OR) symbols.

```
library(dplyr)

# select females
newdata <- filter(starwars,
                  gender == "female")

# select females that are from Alderaan
newdata <- select(starwars,
                  gender == "female" &
                  homeworld == "Alderaan")


# select individuals that are from Alderaan, Coruscant, or Endor
newdata <- select(starwars,
                  homeworld == "Alderaan" |
                  homeworld == "Coruscant" |
                  homeworld == "Endor")

# this can be written more succinctly as
newdata <- select(starwars,
                  homeworld %in%
                    c("Alderaan", "Coruscant", "Endor"))
```

2.2.3   Creating/Recoding Variables

The `mutate` function allows you to create new variables or transform existing ones.

```
library(dplyr)

# convert height in centimeters to inches,
# and mass in kilograms to pounds
newdata <- mutate(starwars,
                  height = height * 0.394,
                  mass   = mass   * 2.205)
```

The `ifelse` function (part of base R) can be used for recoding data. The format is `ifelse(test, return if TRUE, return if FALSE)`.

```
library(dplyr)

# if height is greater than 180 then heightcat = "tall",
# otherwise heightcat = "short"

newdata <- mutate(starwars,
                  heightcat = ifelse(height > 180,
                                     "tall",
                                     "short")

# convert any eye color that is not black, blue or brown, to other.
newdata <- mutate(starwars,
                  eye_color = ifelse(eye_color %in%
                                     c("black", "blue", "brown"),
                                     eye_color,
                                     "other"))

# set heights greater than 200 or less than 75 to missing
newdata <- mutate(starwars,
                  height = ifelse(height < 75 | height > 200,
                                  NA,
                                  height))
```

2.2.4   Summarizing Data

The summarize function can be used to reduce multiple values down to a single value (such as a mean). It is often used in conjunction with the by_group function, to calculate statistics by group. In the code below, the na.rm=TRUE option is used to drop missing values before calculating the means.

```
library(dplyr)

# calculate mean height and mass
newdata <- summarize(starwars,
                     mean_ht = mean(height, na.rm=TRUE),
                     mean_mass = mean(mass, na.rm=TRUE))
newdata

## # A tibble: 1 x 2
##   mean_ht mean_mass
##     <dbl>     <dbl>
## 1    174.      97.3

# calculate mean height and weight by gender
newdata <- group_by(starwars, gender)
newdata <- summarize(newdata,
```

```
                            mean_ht = mean(height, na.rm=TRUE),
                            mean_wt = mean(mass, na.rm=TRUE))
newdata
```

```
## # A tibble: 3 x 3
##    gender      mean_ht mean_wt
##    <chr>         <dbl>   <dbl>
## 1 feminine       165.    54.7
## 2 masculine      177.    106.
## 3 <NA>           181.    48
```

Graphs are often created from summarized data, rather than from the original observations. You will see several examples in Chapter 4.

2.2.5   Using Pipes

Packages like **dplyr** and **tidyr** allow you to write your code in a compact format using the pipe %>% operator. Here is an example.

```
library(dplyr)
```

```
# calculate the mean height for women by species
newdata <- filter(starwars,
                  gender == "female")
newdata <- group_by(species)
newdata <- summarize(newdata,
                     mean_ht = mean(height, na.rm = TRUE))
```

```
# this can be written as more succinctly as
newdata <- starwars %>%
  filter(gender == "female") %>%
  group_by(species) %>%
  summarize(mean_ht = mean(height, na.rm = TRUE))
```

The %>% operator passes the result on the left to the first parameter of the function on the right.

2.2.6   Processing Dates

Date values are entered in R as character values. For example, consider the following simple dataset recording the birth date of three individuals.

```
df <- data.frame(
  dob = c("11/10/1963", "Jan-23-91", "12:1:2001")
)
```

TABLE 2.2
Wide Data

ID	Name	Sex	Height	Weight
01	Bill	Male	70	180
02	Bob	Male	72	195
03	Mary	Female	62	130

```
# view struction of data frame
str(df)

## 'data.frame':    3 obs. of  1 variable:
##  $ dob: chr  "11/10/1963" "Jan-23-91" "12:1:2001"
```

There are many ways to convert character variables to *Date* variables. One of the simplest is to use the functions provided in the **lubridate** package. These include `ymd`, `dmy`, and `mdy` for importing year-month-day, day-month-year, and month-day-year formats, respectively.

```
library(lubridate)
# convert dob from character to date
df$dob <- mdy(df$dob)
str(df)

## 'data.frame':    3 obs. of  1 variable:
##  $ dob: Date, format: "1963-11-10" "1991-01-23" ...
```

The values are recorded internally as the number of days since January 1, 1970. Now that the variable is a Date variable, you can perform date arithmetic (how old are they now), extract date elements (month, day, year), and reformat the values (e.g., October 11, 1963). Date variables are important for time-dependent graphs (Chapter 8).

2.2.7   Reshaping Data

Some graphs require the data to be in wide format, while some graphs require the data to be in long format. An example of wide data is given in Table 2.2. The long format of this data is given in Table 2.3.

You can convert a wide dataset to a long dataset using

```
# convert wide dataset to long dataset
library(tidyr)
long_data <- pivot_longer(wide_data,
                          cols = c("height", "weight"),
                          names_to = "variable",
                          values_to ="value")
```

TABLE 2.3
Long Data

ID	Name	Sex	Variable	Value
01	Bill	Male	Height	70
01	Bill	Male	Weight	180
02	Bob	Male	Height	72
02	Bob	Male	Weight	195
03	Mary	Female	Height	62
03	Mary	Female	Weight	130

Conversely, you can convert a long dataset to a wide dataset using

```
# convert long dataset to wide dataset
library(tidyr)
wide_data <- pivot_wider(long_data,
                        names_from = "variable",
                        values_from = "value")
```

2.2.8   Missing Data

Real data is likely to contain missing values. There are three basic approaches to dealing with missing data: feature selection, listwise deletion, and imputation. Let's see how each applies to the `msleep` dataset from the **ggplot2** package. The msleep dataset describes the sleep habits of mammals and contains missing values on several variables. (See Appendix A.3.)

2.2.8.1   Feature Selection

In feature selection, you delete variables (columns) that contain too many missing values.

```
data(msleep, package="ggplot2")

# what is the proportion of missing data for each variable?
pctmiss <- colSums(is.na(msleep))/nrow(msleep)
round(pctmiss, 2)
```

```
## name genus vore order  conservation sleep_total
## 0.00  0.00 0.08 0.00          0.35        0.00
## sleep_rem sleep_cycle awake brainwt bodywt
##      0.27        0.61  0.00    0.33   0.00
```

Sixty-two percent of the sleep_cycle values are missing. You may decide to drop it.

2.2.8.2 Listwise Deletion

Listwise deletion involves deleting observations (rows) that contain missing values on *any* of the variables of interest.

```
# Create a dataset containing genus, vore, and conservation.
# Delete any rows containing missing data.
newdata <- select(msleep, genus, vore, conservation)
newdata <- na.omit(newdata)
```

2.2.8.3 Imputation

Imputation involves replacing missing values with "reasonable" guesses about what the values would have been if they had not been missing. There are several approaches, as detailed in such packages as **VIM**, **mice**, **Amelia**, and **missForest**. Here, we will use the kNN() function from the **VIM** package to replace missing values with imputed values.

```
# Impute missing values using the 5 nearest neighbors
library(VIM)
newdata <- kNN(msleep, k=5)
```

Basically, for each case with a missing value, the k most similar cases not having a missing value are selected. If the missing value is numeric, the median of those k cases is used as the imputed value. If the missing value is categorical, the most frequent value from the k cases is used. The process iterates over cases and variables until the results converge (become stable). This is a bit of an oversimplification–see Kowarik and Templ (2016) for the actual details.

Important caveat: Missing values can bias the results of studies (sometimes severely). If you have a significant amount of missing data, it is probably a good idea to consult a statistician or data scientist before deleting cases or imputing missing values.

3

Introduction to ggplot2

This chapter provides a brief overview of how the **ggplot2** package works. It introduces the central concepts used to develop an informative graph by exploring the relationships contained in the insurance dataset.

3.1 A Worked Example

The functions in the **ggplot2** package build up a graph in layers. We'll build a complex graph by starting with a simple graph and adding additional elements, one at a time.

The example explores the relationship between smoking, obesity, age, and medical costs using data from the Medical Insurance Costs dataset (Appendix A.4).

First, lets import the data.

```
# load the data
url <- "https://tinyurl.com/mtktm8e5"
insurance <- read.csv(url)
```

Next, we'll add a variable indicating if the patient is obese or not. Obesity will be defined as a body mass index greater than or equal to 30.

```
# create an obesity variable
insurance$obese <- ifelse(insurance$bmi >= 30,
                          "obese", "not obese")
```

In building a **ggplot2** graph, only the first two functions described below are required. The others are optional and can appear in any order.

3.1.1 ggplot

The first function in building a graph is the `ggplot` function. It specifies the data frame to be used and the mapping of the variables to the visual properties of the graph. The mappings are placed within the `aes` function,

DOI: 10.1201/9781003299271-3

FIGURE 3.1
Map variables.

which stands for aesthetics. Let's start by looking at the relationship between age and medical expenses.

```
# specify dataset and mapping
library(ggplot2)
ggplot(data = insurance,
       mapping = aes(x = age, y = expenses))
```

Why is the graph in Figure 3.1 empty? We specified that the *age* variable should be mapped to the *x*-axis and that the *expenses* should be mapped to the *y*-axis, but we haven't yet specified what we wanted placed on the graph.

3.1.2 geoms

Geoms are the geometric objects (points, lines, bars, etc.) that can be placed on a graph. They are added using functions that start with geom_. In this example, we'll add points using the geom_point function, creating a scatterplot.

In **ggplot2** graphs, functions are chained together using the + sign to build a final plot.

```
# add points
ggplot(data = insurance,
       mapping = aes(x = age, y = expenses)) +
  geom_point()
```

Figure 3.2 indicates that expenses rise with age in a fairly linear fashion.

A number of parameters (options) can be specified in a `geom_` function. Options for the `geom_point` function include `color`, `size`, and `alpha`. These control the point color, size, and transparency, respectively. Transparency ranges from 0 (completely transparent) to 1 (completely opaque). Adding a degree of transparency can help visualize overlapping points. In Figure 3.3, the points are light blue, slightly larger, and semi-transparent.

```
# make points blue, larger, and semi-transparent
ggplot(data = insurance,
       mapping = aes(x = age, y = expenses)) +
  geom_point(color = "cornflowerblue",
             alpha = .7,
             size = 2)
```

Next, let's add a line of best fit (Figure 3.4). We can do this with the `geom_smooth` function. Options control the type of line (linear, quadratic,

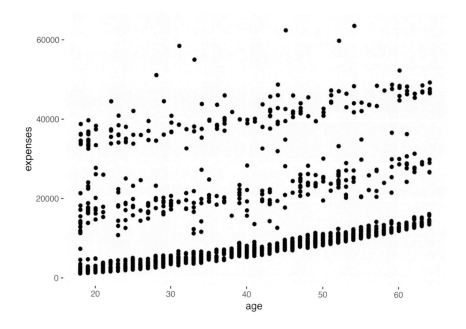

FIGURE 3.2
Add points.

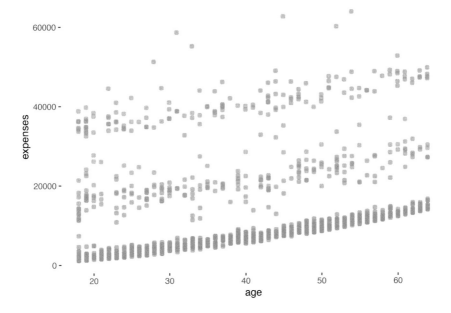

FIGURE 3.3
Modify point color, transparency, and size.

nonparametric), the thickness of the line, the line's color, and the presence or absence of a confidence interval. Here, we request a linear regression (`method = lm`) line (where *lm* stands for linear model).

```
# add a line of best fit.
ggplot(data = insurance,
       mapping = aes(x = age, y = expenses)) +
  geom_point(color = "cornflowerblue",
             alpha = .5,
             size = 2) +
  geom_smooth(method = "lm")
```

Expenses appear to increase with age, but there is an unusual clustering of the point. We will find out why as we delve deeper into the data.

3.1.3 grouping

In addition to mapping variables to the *x* and *y* axes, variables can be mapped to the color, shape, size, transparency, and other visual characteristics of geometric objects. This allows groups of observations to be superimposed in a single graph.

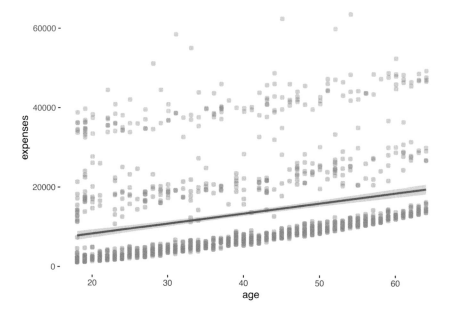

FIGURE 3.4
Add line of best fit.

Let's add smoker status to the plot and represent it by color.

```
# indicate sex using color
ggplot(data = insurance,
       mapping = aes(x = age,
                     y = expenses,
                     color = smoker)) +
  geom_point(alpha = .5,
             size = 2) +
  geom_smooth(method = "lm",
              se = FALSE,
              size = 1.5)
```

The `color = smoker` option is place in the `aes` function, because we are mapping a variable to an aesthetic (a visual characteristic of the graph). The `geom_smooth` option (`se = FALSE`) was added to suppresses the confidence intervals. The resulting graph is given in Figure 3.5.

It appears that smokers tend to incur greater expenses than non-smokers (not a surprise).

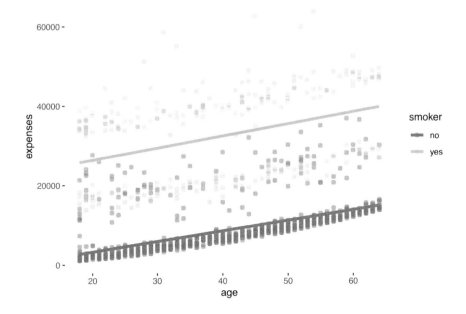

FIGURE 3.5
Include sex, using color.

3.1.4   scales

Scales control how variables are mapped to the visual characteristics of the plot. Scale functions (which start with `scale_`) allow you to modify this mapping. In the next plot (Figure 3.6), we'll change the x- and y-axis scaling, and the colors employed.

```
# modify the x and y axes and specify the colors to be used
ggplot(data = insurance,
       mapping = aes(x = age,
                     y = expenses,
                     color = smoker)) +
  geom_point(alpha = .5,
             size = 2) +
  geom_smooth(method = "lm",
              se = FALSE,
              size = 1.5) +
  scale_x_continuous(breaks = seq(0, 70, 10)) +
  scale_y_continuous(breaks = seq(0, 60000, 20000),
                     label = scales::dollar) +
  scale_color_manual(values = c("indianred3",
                                "cornflowerblue"))
```

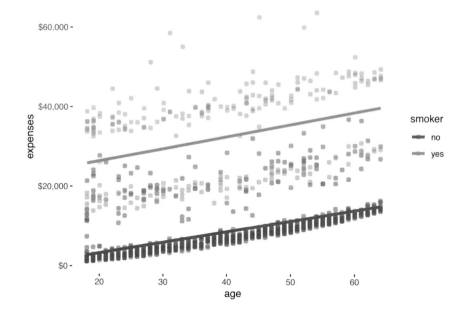

FIGURE 3.6
Change colors and axis labels.

We're getting there. Here is a question. Is the relationship between age, expenses, and smoking the same for obese and non-obese patients? Let's repeat this graph once for each weight status in order to explore this.

3.1.5   facets

Facets reproduce a graph for each level a given variable (or pair of variables). Facets are created using functions that start with `facet_`. Here, facets will be defined by the two levels of the *obese* variable.

```
# reproduce plot for each obsese and non-obese individuals
ggplot(data = insurance,
       mapping = aes(x = age,
                     y = expenses,
                     color = smoker)) +
  geom_point(alpha = .5) +
  geom_smooth(method = "lm",
              se = FALSE) +
  scale_x_continuous(breaks = seq(0, 70, 10)) +
  scale_y_continuous(breaks = seq(0, 60000, 20000),
                     label = scales::dollar) +
  scale_color_manual(values = c("indianred3",
```

```
                                    "cornflowerblue")) +
facet_wrap(~obese)
```

From Figure 3.7 we can simultaneously visualize the relationships among age, smoking status, obesity, and annual medical expenses.

3.1.6   labels

Graphs should be easy to interpret, and informative labels are a key element in achieving this goal. The labs function provides customized labels for the axes and legends. Additionally, a custom title, subtitle, and caption can be added (Figure 3.8).

```
# add informative labels
ggplot(data = insurance,
       mapping = aes(x = age,
                     y = expenses,
                     color = smoker)) +
  geom_point(alpha = .5) +
  geom_smooth(method = "lm",
              se = FALSE) +
  scale_x_continuous(breaks = seq(0, 70, 10)) +
```

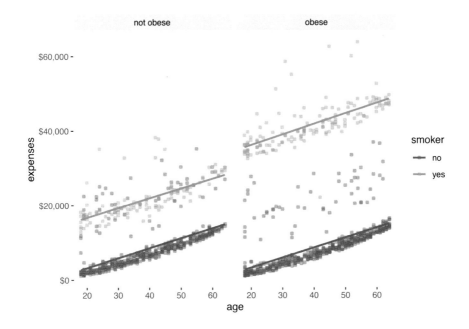

FIGURE 3.7
Add job sector, using faceting.

```
scale_y_continuous(breaks = seq(0, 60000, 20000),
                   label = scales::dollar) +
scale_color_manual(values = c("indianred3",
                              "cornflowerblue")) +
facet_wrap(~obese) +
labs(title = "Relationship between patient demographics and medical
             costs",
     subtitle = "US Census Bureau 2013",
     caption = "source: http://mosaic-web.org/",
     x = " Age (years)",
     y = "Annual expenses",
     color = "Smoker?")
```

Now a viewer doesn't need to guess what the labels *expenses* and *age* mean or where the data come from.

3.1.7 themes

Finally, we can fine-tune the appearance of the graph using themes. Theme functions (which start with `theme_`) control background colors, fonts, grid-lines, legend placement, and other non-data related features of the graph. Let's use a cleaner theme.

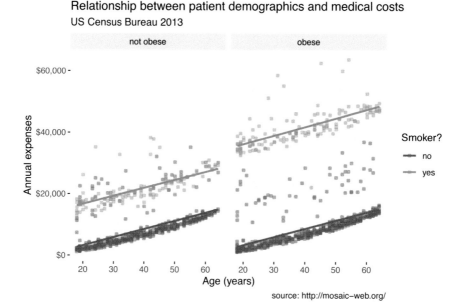

FIGURE 3.8
Add informative titles and labels.

```
# use a minimalist theme
ggplot(data = insurance,
       mapping = aes(x = age,
                     y = expenses,
                     color = smoker)) +
  geom_point(alpha = .5) +
  geom_smooth(method = "lm",
              se = FALSE) +
  scale_x_continuous(breaks = seq(0, 70, 10)) +
  scale_y_continuous(breaks = seq(0, 60000, 20000),
                     label = scales::dollar) +
  scale_color_manual(values = c("indianred3",
                                "cornflowerblue")) +
  facet_wrap(~obese) +
  labs(title = "Relationship between age and medical expenses",
       subtitle = "US Census Data 2013",
       caption = "source: https://github.com/dataspelunking/MLwR",
       x = " Age (years)",
       y = "Medical Expenses",
       color = "Smoker?") +
  theme_minimal()
```

Now, we have something. From Figure 3.9 it appears that:

- There is a positive linear relationship between age and expenses. The relationship is constant across smoking and obesity status (i.e., the slope doesn't change).
- Smokers and obese patients have higher medical expenses.
- There is an interaction between smoking and obesity. Non-smokers look fairly similar across obesity groups. However, for smokers, obese patients have much higher expenses.
- There are some very high outliers (large expenses) among the obese smoker group.

These findings are tentative. They are based on a limited sample size and do not involve statistical testing to assess whether differences may be due to chance variation.

3.2   Placing the `data` and `mapping` Options

Plots created with **ggplot2** always start with the `ggplot` function. In the examples above, the `data` and `mapping` options were placed in this function. In this case they apply to each `geom_` function that follows. You can also place these options directly within a `geom`. In that case, they apply only to that specific geom.

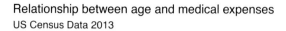

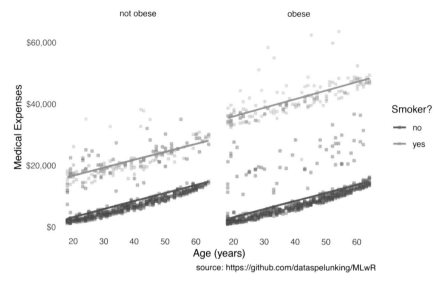

FIGURE 3.9
Use a simpler theme.

Consider the following graph.

```
# placing color mapping in the ggplot function
ggplot(insurance,
       aes(x = age,
           y = expenses,
           color = smoker)) +
  geom_point(alpha = .5,
             size = 2) +
  geom_smooth(method = "lm",
              se = FALSE,
              size = 1.5)
```

Since the mapping of the variable smoker to color appears in the `ggplot` function, it applies to *both* `geom_point` and `geom_smooth`. The point color indicates the smoker status, and a separate colored trend line is produced for smokers and non-smokers (Figure 3.10). Compare this to Figure 3.11, created by the following code.

```
# placing color mapping in the geom_point function
ggplot(insurance,
       aes(x = age,
```

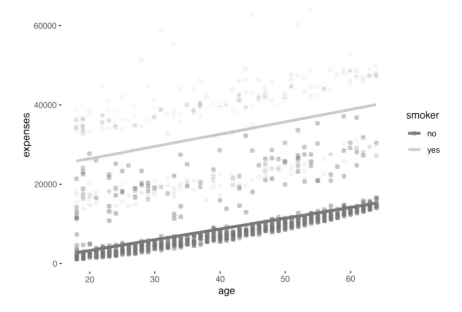

FIGURE 3.10
Color mapping in ggplot function.

```
          y = expenses)) +
  geom_point(aes(color = smoker),
             alpha = .5,
             size = 2) +
  geom_smooth(method = "lm",
              se = FALSE,
              size = 1.5)
```

Since the smoker to color mapping only appears in the `geom_point` function, it is only used there. A single trend line is created for all observations (Figure 3.11).

Most of the examples in this book place the data and mapping options in the `ggplot` function. Additionally, the phrases *data=* and *mapping=* are omitted since the first option always refers to data and the second option always refers to mapping.

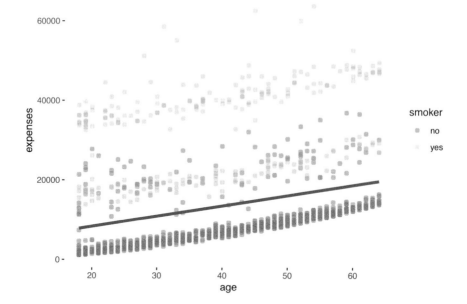

FIGURE 3.11
Color mapping in geom_point function.

3.3   Graphs as Objects

A **ggplot2** graph can be saved as a named R object (like a data frame),
manipulated further, and then printed or saved to disk.

```
# create scatterplot and save it
myplot <- ggplot(data = insurance,
                 aes(x = age, y = expenses)) +
            geom_point()

# plot the graph
myplot

# make the points larger and blue
# then print the graph
myplot <- myplot + geom_point(size = 2, color = "blue")
myplot

# print the graph with a title and line of best fit
# but don't save those changes
```

```
myplot + geom_smooth(method = "lm") +
  labs(title = "Mildly interesting graph")

# print the graph with a black and white theme
# but don't save those changes
myplot + theme_bw()
```

This can be a real time saver (and help you avoid carpal tunnel syndrome). It is also handy when saving graphs programmatically.

Now it's time to apply what we've learned.

4

Univariate Graphs

The first step in any comprehensive data analysis is to explore each import variable in turn. Univariate graphs plot the distribution of data from a single variable. The variable can be categorical (e.g., race, sex, political affiliation) or quantitative (e.g., age, weight, income).

The dataset `Marriage` contains the marriage records of 98 individuals in Mobile County, Alabama (see Appendix A.5). We'll explore the distribution of three variables from this dataset–the age and race of the wedding participants, and the occupation of the wedding officials.

4.1 Categorical

The race of the participants and the occupation of the officials are both categorical variables.The distribution of a single categorical variable is typically plotted with a bar chart, a pie chart, or (less commonly) a tree map or waffle chart.

4.1.1 Bar Chart

In Figure 4.1, a bar chart is used to display the distribution of wedding participants by race.

```
# simple bar chart
library(ggplot2)
data(Marriage, package = "mosaicData")

# plot the distribution of race
ggplot(Marriage, aes(x = race)) +
  geom_bar()
```

The majority of participants are white, followed by black, with very few Hispanics or American Indians.

You can modify the bar fill and border colors, plot labels, and title by adding options to the `geom_bar` function. In **ggplot2**, the `fill` parameter is

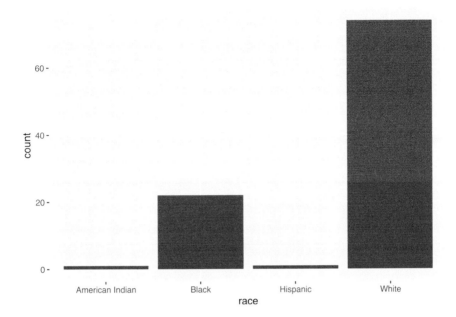

FIGURE 4.1
Simple barchart.

used to specify the color of areas such as bars, rectangles, and polygons. The
`color` parameter specifies the color objects that technically do not have an
area, such as points, lines, and borders. In Figure 4.2, the bars are cornflower
blue, with black borders. Additionally, a plot title and axis labels have been
added.

```
# plot the distribution of race with modified colors and labels
ggplot(Marriage, aes(x=race)) +
  geom_bar(fill = "cornflowerblue",
           color="black") +
  labs(x = "Race",
       y = "Frequency",
       title = "Participants by race")
```

4.1.1.1 Percents

Bars can represent percents rather than counts. For bar charts, the
code `aes(x=race)` is actually a shortcut for `aes(x = race, y =
after_stat(count))`, where `count` is a special variable representing
the frequency within each category. You can use this to calculate percentages,
by specifying `y` variable explicitly.

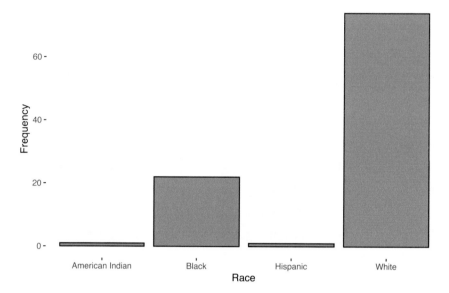

FIGURE 4.2
Barchart with modified colors, labels, and title.

```
# plot the distribution as percentages
ggplot(Marriage,
       aes(x = race, y = after_stat(count/sum(count)))) +
  geom_bar() +
  labs(x = "Race",
       y = "Percent",
       title = "Participants by race") +
  scale_y_continuous(labels = scales::percent)
```

In the code above, the `scales` package is used to add % symbols to the *y*-axis labels. The results are given in Figure 4.3.

4.1.1.2   Sorting Categories

It is often helpful to sort the bars by frequency. In the code below, the frequencies are calculated explicitly. Then the `reorder` function is used to sort the categories by the frequency. The option `stat="identity"` tells the plotting function not to calculate counts, because they are supplied directly.

```
# calculate number of participants in each race category
library(dplyr)
```

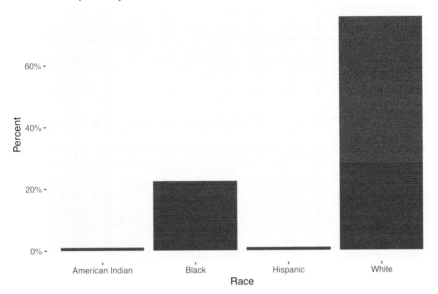

FIGURE 4.3
Barchart with percentages.

```
plotdata <- Marriage %>%
 count(race)
```

The resulting dataset is give in Table 4.1.
This new dataset is then used to create the graph in Figure 4.4.

```
# plot the bars in ascending order
ggplot(plotdata,
       aes(x = reorder(race, n), y = n)) +
  geom_bar(stat="identity") +
  labs(x = "Race",
       y = "Frequency",
       title  = "Participants by race")
```

TABLE 4.1
plotdata

Race	n
American Indian	1
Black	22
Hispanic	1
White	74

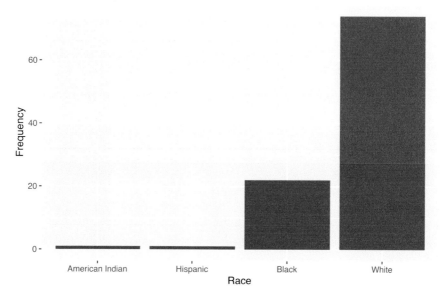

FIGURE 4.4
Sorted bar chart.

The graph bars are sorted in ascending order. Use `reorder(race, -n)` to sort in descending order.

4.1.1.3 Labeling Bars

Finally, you may want to label each bar with its numerical value (Figure 4.5).

```
# plot the bars with numeric labels
ggplot(plotdata,
       aes(x = race, y = n)) +
  geom_bar(stat="identity") +
  geom_text(aes(label = n), vjust=-0.5) +
  labs(x = "Race",
       y = "Frequency",
       title = "Participants by race")
```

Here `geom_text` adds the labels, and `vjust` controls vertical justification. See Annotations (Section 11.7) for more details.

Putting these ideas together, you can create a graph like the one in Figure 4.6. The minus sign in `reorder(race, -pct)` is used to order the bars in descending order.

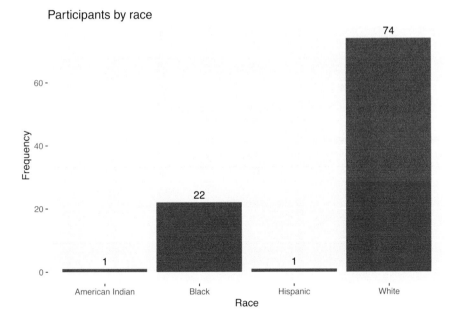

FIGURE 4.5
Bar chart with numeric labels.

```
library(dplyr)
library(scales)
plotdata <- Marriage %>%
  count(race) %>%
  mutate(pct = n / sum(n),
         pctlabel = paste0(round(pct*100), "%"))

# plot the bars as percentages,
# in decending order with bar labels
ggplot(plotdata,
       aes(x = reorder(race, -pct), y = pct)) +
  geom_bar(stat="identity", fill="indianred3", color="black") +
  geom_text(aes(label = pctlabel), vjust=-0.25) +
  scale_y_continuous(labels = percent) +
  labs(x = "Race",
       y = "Percent",
       title  = "Participants by race")
```

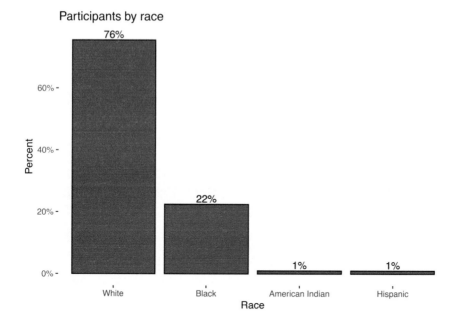

FIGURE 4.6
Sorted bar chart with percent labels.

4.1.1.4   Overlapping Labels

Category labels may overlap if (1) there are many categories or (2) the labels
are long. Consider the distribution of marriage officials. In Figure 4.7, the
x-axis labels are unreadable.

```
# basic bar chart with overlapping labels
ggplot(Marriage, aes(x=officialTitle)) +
  geom_bar() +
  labs(x = "Officiate",
       y = "Frequency",
       title = "Marriages by officiate")
```

In this case, you can flip the x and y axes with the `coord_flip` function
(Figure 4.8).

```
# horizontal bar chart
ggplot(Marriage, aes(x = officialTitle)) +
  geom_bar() +
  labs(x = "",
       y = "Frequency",
       title = "Marriages by officiate") +
  coord_flip()
```

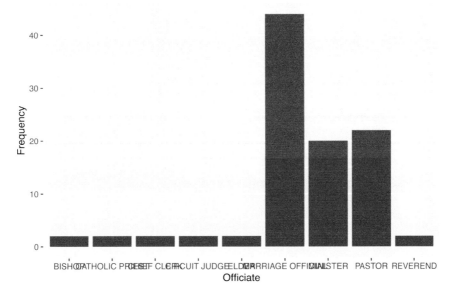

FIGURE 4.7
Barchart with problematic labels.

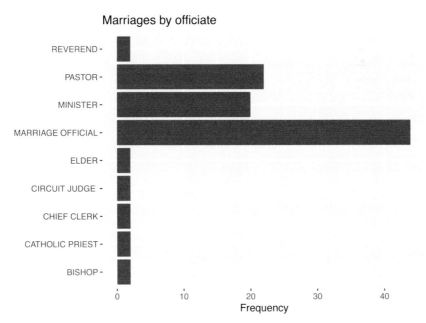

FIGURE 4.8
Horizontal barchart.

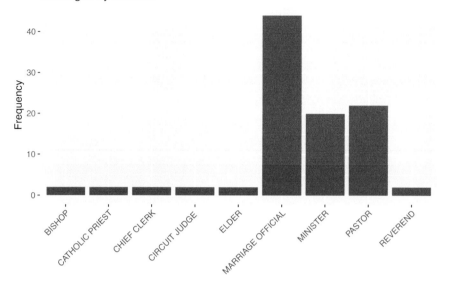

FIGURE 4.9
Barchart with rotated labels.

Alternatively, you can rotate the axis labels (Figure 4.9).

```
# bar chart with rotated labels
ggplot(Marriage, aes(x=officialTitle)) +
  geom_bar() +
  labs(x = "",
       y = "Frequency",
       title = "Marriages by officiate") +
  theme(axis.text.x = element_text(angle = 45,
                                    hjust = 1))
```

Finally, you can try staggering the labels. The trick is to add a newline \n to every other label (Figure 4.10).

```
# bar chart with staggered labels
lbls <- paste0(c("","\n"), levels(Marriage$officialTitle))
ggplot(Marriage,
       aes(x=factor(officialTitle,
                    labels = lbls))) +
  geom_bar() +
  labs(x = "",
       y = "Frequency",
       title = "Marriages by officiate")
```

In general, I recommend trying not to rotate axis labels. It places a greater cognitive demand on the end user (i.e., it is harder to read!).

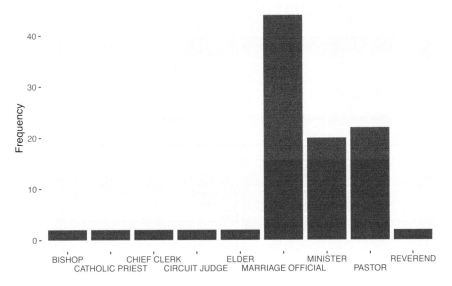

Marriages by officiate

FIGURE 4.10
Barchart with staggered labels.

4.1.2 Pie Chart

Pie charts are controversial in statistics. If your goal is to compare the frequency of categories, you are better off with bar charts (humans are better at judging the length of bars than the volume of pie slices). If your goal is to compare each category with the the whole (e.g., what portion of participants are Hispanic compared to all participants), and the number of categories is small, then pie charts may work for you.

Pie charts are easily created with `ggpie` function in the **ggpie** package. The format is `ggpie(data, variable)`, where *data* is a data frame, and *variable* is the categorical variable to be plotted. A basic pie chart is given in Figure 4.11.

```
# create a basic ggplot2 pie chart
library(ggpie)
ggpie(Marriage, race)
```

The `ggpie` function has many option, as described in package homepage (http://rkabacoff.github.io/ggpie). For example to place the labels within the pie, set `legend = FALSE`. A title can be added with the `title` option. The revised pie chart is given in Figure 4.12. Placing the labels within the slices makes the graph easier to read.

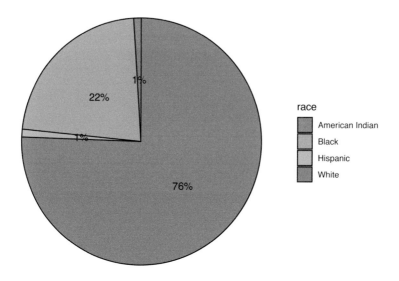

FIGURE 4.11
Basic pie chart with legend.

```
# create a pie chart with slice labels within figure
ggpie(Marriage, race, legend = FALSE, title = "Participants by race")
```

The pie chart makes it easy to compare each slice with the whole. For example, roughly a quarter of the total participants are Black.

4.1.3   Tree Map

An alternative to a pie chart is a tree map. Unlike pie charts, it can handle categorical variables that have *many* levels. The code below creates the graph in Figure 4.13.

```
library(treemapify)

# create a treemap of marriage officials
plotdata <- Marriage %>%
  count(officialTitle)

ggplot(plotdata,
       aes(fill = officialTitle, area = n)) +
  geom_treemap() +
  labs(title = "Marriages by officiate")
```

Here is a more useful version with the labels placed within the tiles. The resulting graph is given in Figure 4.14.

Participants by race

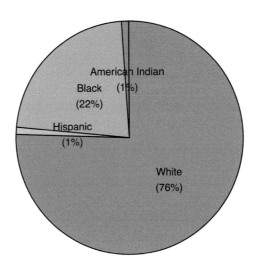

FIGURE 4.12
Pie chart with percent labels.

Marriages by officiate

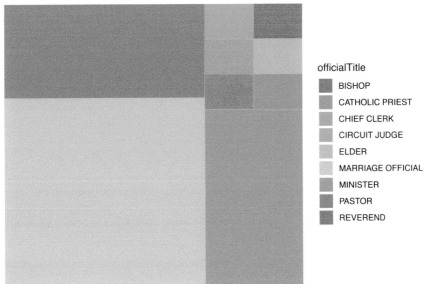

FIGURE 4.13
Basic treemap.

Marriages by officiate

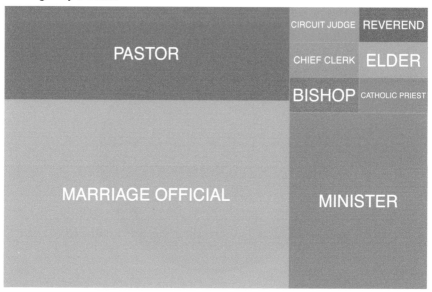

FIGURE 4.14
Treemap with labels.

```
# create a treemap with tile labels
ggplot(plotdata,
       aes(fill = officialTitle,
           area = n,
           label = officialTitle)) +
  geom_treemap() +
  geom_treemap_text(colour = "white",
                    place = "centre") +
  labs(title = "Marriages by officiate") +
  theme(legend.position = "none")
```

The treemapify package offers many options for customization. See https://wilkox.org/treemapify/ for details.

4.1.4   Waffle Chart

A waffle chart, also known as a gridplot or square pie chart, represents observations as squares in a rectangular grid, where each cell represents a percentage of the whole. You can create a **ggplot2** waffle chart using the `geom_waffle` function in the **waffle** package.

Let's create a waffle chart for the professions of wedding officiates. As with tree maps, start by summarizing the data into groups and counts.

```
library(dplyr)
plotdata <- Marriage %>%
  count(officialTitle)
```

Next create the **ggplot2** graph (Figure 4.14). Set *fill* to the grouping variable and *values* to the counts. Don't specify an *x* and *y*.

Note: The na.rm parameter in the geom_waffle function indicates whether missing values should be deleted. At time of this writing, there is a bug in the function. The default for the na.rm parameter is NA, but it actually must be either TRUE or FALSE. Specifying one or the other eliminates the error.

The following code produces the default waffle plot (Figure 4.15).

```
# create a basic waffle chart
library(waffle)
ggplot(plotdata, aes(fill = officialTitle, values=n)) +
  geom_waffle(na.rm=TRUE)
```

Next, we'll customize the graph by

- specifying the number of rows and cell sizes and setting borders around the cells to "white" (geom_waffle)
- change the color scheme to "Spectral" (scale_fill_brewer)
- assure that the cells are squares and not rectangles (coord_equal)
- simplify the theme (the theme functions)
- modify the title and add a caption with the scale (labs)

```
# Create a customized caption
cap <- paste0("1 square = ", ceiling(sum(plotdata$n)/100),
              " case(s).")
library(waffle)
ggplot(plotdata, aes(fill = officialTitle, values=n)) +
  geom_waffle(na.rm=TRUE,
              n_rows = 10,
              size = .4,
              color = "white") +
  scale_fill_brewer(palette = "Spectral") +
  coord_equal() +
  theme_minimal() +
  theme_enhance_waffle() +
  theme(legend.title = element_blank()) +
```

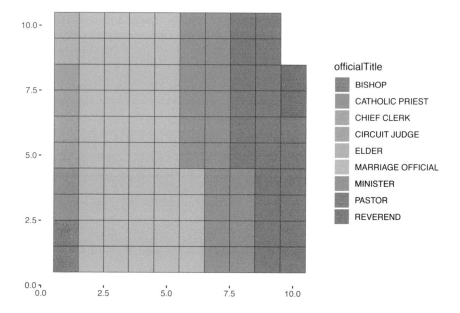

FIGURE 4.15
Basic waffle chart.

```
labs(title = "Proportion of Wedding Officials",
     caption = cap)
```

The customized graph is presented in Figure 4.16. While new to R, waffle charts are becoming increasingly popular.

4.2 Quantitative

In the `Marriage` dataset, age is quantitative variable. The distribution of a single quantitative variable is typically plotted with a histogram, kernel density plot, or dot plot.

4.2.1 Histogram

Histograms are the most common approach to visualizing a quantitative variable. In a histogram, the values of a variable are typically divided up into adjacent, equal width ranges (called *bins*), and the number of observations in each bin is plotted with a vertical bar.

Proportion of Wedding Officials

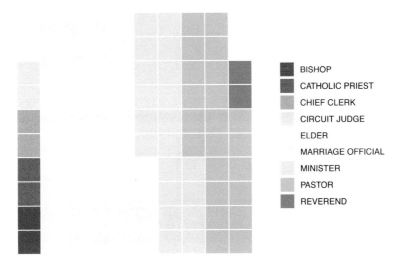

1 square = 1 case(s).

FIGURE 4.16

Customized waffle chart.

The following code creates a basic histogram (Figure 4.17). By default, the observations are divided into 30 bins.

```
library(ggplot2)

# plot the age distribution using a histogram
ggplot(Marriage, aes(x = age)) +
  geom_histogram() +
  labs(title = "Participants by age",
       x = "Age")
```

Most participants appear to be in their early 20s with another group in their 40's, and a much smaller group in their late 60s and early 70s. This would be a *multimodal* distribution.

Histogram colors can be modified using two options:

- `fill`–fill color for the bars
- `color`–border color around the bars

Here is a histogram using light blue bars with white borders (Figure 4.18).

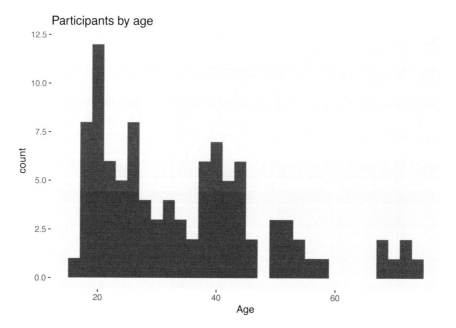

FIGURE 4.17
Basic histogram.

```
# plot the histogram with blue bars and white borders
ggplot(Marriage, aes(x = age)) +
  geom_histogram(fill = "cornflowerblue",
                 color = "white") +
  labs(title="Participants by age",
       x = "Age")
```

4.2.1.1   Bins and Bandwidths

One of the most important histogram options is `bins`, which controls the number of bins into which the numeric variable is divided (i.e., the number of bars in the plot). The default is 30, but it is helpful to try smaller and larger numbers to get a better impression of the shape of the distribution. Figure 4.19 displays the same data with 20 bins.

```
# plot the histogram with 20 bins
ggplot(Marriage, aes(x = age)) +
  geom_histogram(fill = "cornflowerblue",
                 color = "white",
                 bins = 20) +
  labs(title="Participants by age",
```

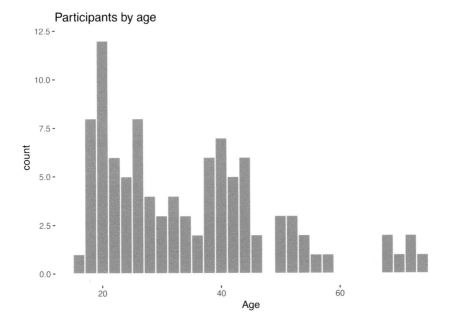

FIGURE 4.18
Histogram with specified fill and border colors.

```
            subtitle = "number of bins = 20",
            x = "Age")
```

Alternatively, you can specify the `binwidth`, the width of the bins represented by the bars. Figure 4.20 displays the data with binwidth that are 5 years wide.

```
# plot the histogram with a binwidth of 5
ggplot(Marriage, aes(x = age)) +
  geom_histogram(fill = "cornflowerblue",
                 color = "white",
                 binwidth = 5) +
  labs(title="Participants by age",
       subtitle = "binwidth = 5 years",
       x = "Age")
```

As with bar charts, the *y*-axis can represent counts or percent of the total (Figure 4.21). Percents are plotted using the code snippet `y = after_stat(count/sum(count))` in the `aes` function. In this case, the height of the bars is not calculated from a variable in the original data. Instead it's calculated from bin counts.

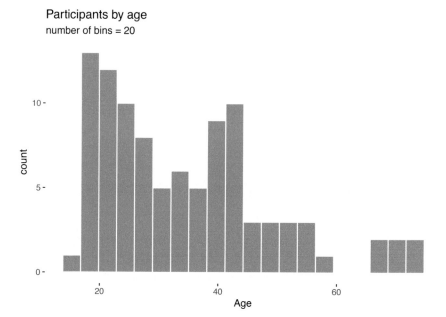

FIGURE 4.19
Histogram with a specified number of bins.

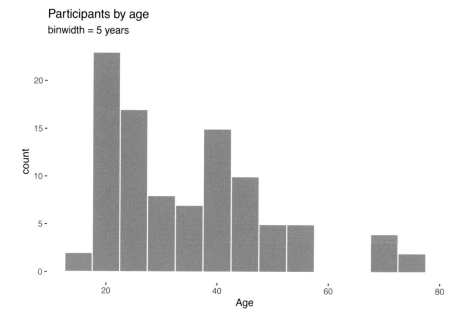

FIGURE 4.20
Histogram with specified a bin width.

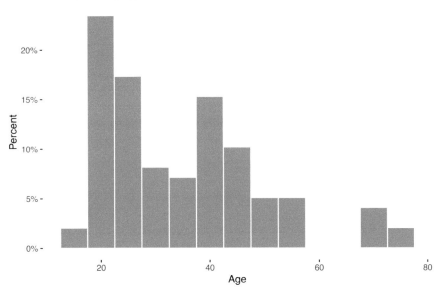

FIGURE 4.21
Histogram with percentages on the *y*-axis.

```
# plot the histogram with percentages on the y-axis
library(scales)
ggplot(Marriage,
       aes(x = age, y = after_stat(count/sum(count)))) +
  geom_histogram(fill = "cornflowerblue",
                 color = "white",
                 binwidth = 5) +
  labs(title="Participants by age",
       y = "Percent",
       x = "Age") +
  scale_y_continuous(labels = percent)
```

4.2.2 Kernel Density Plot

An alternative to a histogram is the kernel density plot. Technically, kernel density estimation is a nonparametric method for estimating the probability density function of a continuous random variable (what??). Basically, we are trying to draw a smoothed histogram, where the area under the curve equals one.

```
# Create a kernel density plot of age
ggplot(Marriage, aes(x = age)) +
  geom_density() +
  labs(title = "Participants by age")
```

The graph (Figure 4.22) shows the distribution of scores. For example, the proportion of cases between 20 and 40 years old would be represented by the area under the curve between 20 and 40 on the x-axis.

As with previous charts, we can use `fill` and `color` to specify the fill and border colors (Figure 4.23).

```
# Create a kernel density plot of age
ggplot(Marriage, aes(x = age)) +
  geom_density(fill = "indianred3") +
  labs(title = "Participants by age")
```

4.2.2.1   Smoothing Parameter

The degree of smoothness is controlled by the bandwidth parameter `bw`. To find the default value for a particular variable, use the `bw.nrd0` function. Values

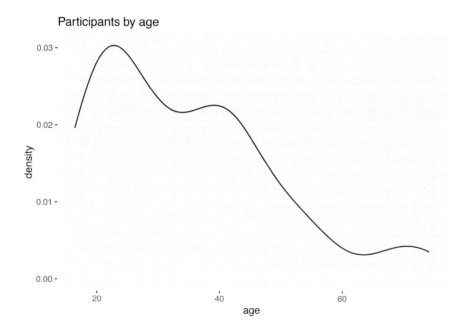

FIGURE 4.22

Basic kernel density plot.

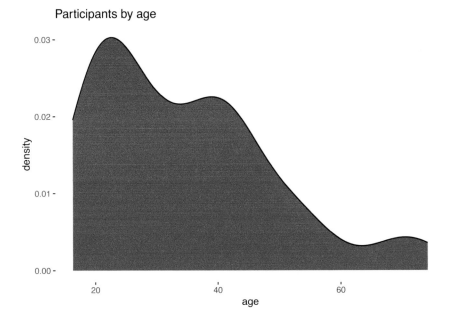

FIGURE 4.23
Kernel density plot with fill.

that are larger will result in more smoothing, while values that are smaller will produce less smoothing.

```
# default bandwidth for the age variable
bw.nrd0(Marriage$age)
```

```
## [1] 5.181946
```

```
# Create a kernel density plot of age
ggplot(Marriage, aes(x = age)) +
  geom_density(fill = "deepskyblue",
               bw = 1) +
  labs(title = "Participants by age",
       subtitle = "bandwidth = 1")
```

In this example, the default bandwidth for age is 5.18. Choosing a value of 1 resulted in less smoothing and more detail (see Figure 4.24).

Kernel density plots allow you to easily see which scores are most frequent and which are relatively rare. However, it can be difficult to explain the meaning of the y-axis means to a non-statistician. (But it will make you look really smart at parties!)

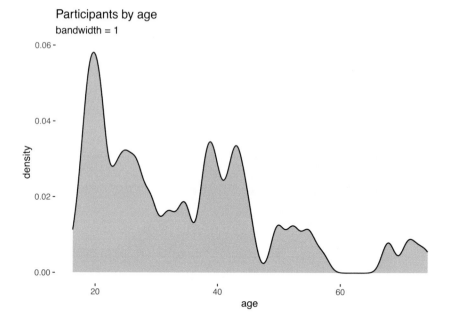

FIGURE 4.24
Kernel density plot with a specified bandwidth.

4.2.3   Dot Chart

Another alternative to the histogram is the dot chart. Again, the quantitative variable is divided into bins, but rather than summary bars, each observation is represented by a dot. By default, the width of a dot corresponds to the bin width, and dots are stacked, with each dot representing one observation. This works best when the number of observations is small (say, less than 150).

The following code creates a basic dot chart (Figure 4.25).

```
# plot the age distribution using a dotplot
ggplot(Marriage, aes(x = age)) +
  geom_dotplot() +
  labs(title = "Participants by age",
       y = "Proportion",
       x = "Age")
```

The `fill` and `color` options can be used to specify the fill and border color of each dot, respectively. In the following code, a dotplot is created with black bordered gold circles (Figure 4.26).

```
# Plot ages as a dot plot using
# gold dots with black borders
ggplot(Marriage, aes(x = age)) +
```

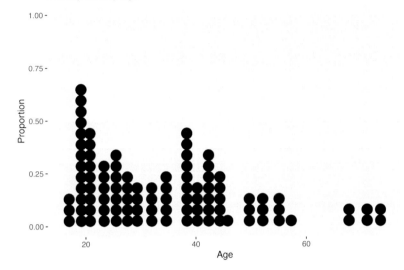

FIGURE 4.25
Basic dotplot.

```
geom_dotplot(fill = "gold",
             color="black") +
```

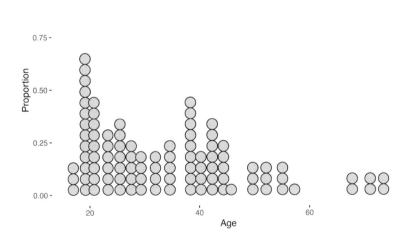

FIGURE 4.26
Dotplot with a specified color scheme.

```
labs(title = "Participants by age",
     y = "Proportion",
     x = "Age")
```

There are many more options available. See `?geom_dotplot` for details and examples.

5

Bivariate Graphs

One of the most fundamental questions in research is *"What is the relationship between A and B?"*. Bivariate graphs display the relationship between two variables. The type of graph will depend on the measurement level of each variable (categorical or quantitative).

5.1 Categorical vs. Categorical

When plotting the relationship between two categorical variables, stacked, grouped, or segmented bar charts are typically used. A less common approach is the mosaic chart (Section 9.5).

In this section, we will look at automobile characteristics contained in mpg dataset that comes with the **ggplot2** package. It provides fuel efficiency data for 38 popular car models in 1998 and 2008 (see Appendix A.6).

5.1.1 Stacked Bar Chart

Let's examine the relationship between automobile class and drive type (front-wheel, rear-wheel, or 4-wheel drive) for the automobiles in the mpg dataset.

```
library(ggplot2)

# stacked bar chart
ggplot(mpg, aes(x = class, fill = drv)) +
  geom_bar(position = "stack")
```

From the Figure 5.1, we can see for example, that the most common vehicle is the SUV. All 2-seater cars are rear wheel drive, while most, but not all SUVs are 4-wheel drive.

Stacked is the default, so the last line could have also been written as `geom_bar()`.

DOI: 10.1201/9781003299271-5   53

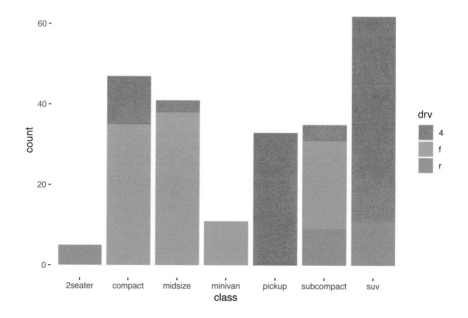

FIGURE 5.1
Stacked bar chart.

5.1.2 Grouped Bar Chart

Grouped bar charts place bars for the second categorical variable side-by-side.
To create a grouped bar plot use the `position = "dodge"` option.

```
library(ggplot2)

# grouped bar plot
ggplot(mpg, aes(x = class, fill = drv)) +
  geom_bar(position = "dodge")
```

From Figure 5.2 you can that all Minivans are front-wheel drive. By default,
zero count bars are dropped and the remaining bars are made wider. This
may not be the behavior you want. You can modify this using the `position
= position_dodge(preserve = "single")"` option.

```
library(ggplot2)

# grouped bar plot preserving zero count bars
ggplot(mpg, aes(x = class, fill = drv)) +
  geom_bar(position = position_dodge(preserve = "single"))
```

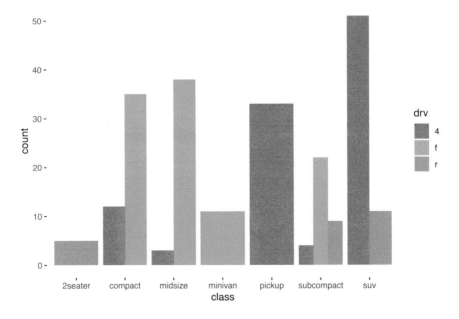

FIGURE 5.2
Side-by-side bar chart.

The results are given in Figure 5.3. Note that this option is only available in the later versions of **ggplot2**.

5.1.3 Segmented Bar Chart

A segmented bar plot is a stacked bar plot where each bar represents 100 percent. You can create a segmented bar chart using the `position = "filled"` option (Figure 5.4).

```
library(ggplot2)

# bar plot, with each bar representing 100%
ggplot(mpg, aes(x = class, fill = drv)) +
  geom_bar(position = "fill") +
  labs(y = "Proportion")
```

This type of plot is particularly useful if the goal is to compare the percentage of a category in one variable across each level of another variable. For example, the proportion of front-wheel drive cars go up as you move from compact, to midsize, to minivan.

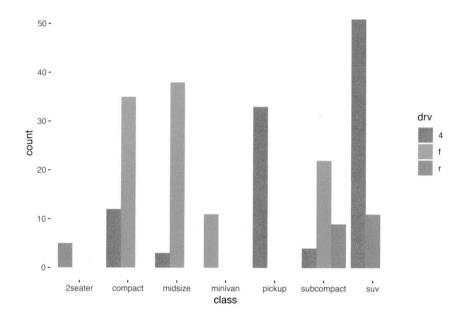

FIGURE 5.3
Side-by-side bar chart with zero count bars retained.

5.1.4   Improving the Color and Labeling

You can use additional options to improve color and labeling. In Figure 5.5

- `factor` modifies the order of the categories for the class variable and both the order and the labels for the drive variable

- `scale_y_continuous` modifies the y-axis tick mark labels

- `labs` provides a title and changed the labels for the x and y axes and the legend
- `scale_fill_brewer` changes the fill color scheme

- `theme_minimal` removes the grey background and changed the grid color

```
library(ggplot2)

# bar plot, with each bar representing 100%,
# reordered bars, and better labels and colors
library(scales)
ggplot(mpg,
       aes(x = factor(class,
```

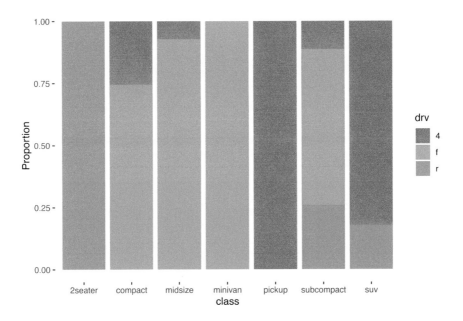

FIGURE 5.4
Segmented bar chart.

```
                    levels = c("2seater", "subcompact",
                               "compact", "midsize",
                               "minivan", "suv", "pickup")),
           fill = factor(drv,
                         levels = c("f", "r", "4"),
                         labels = c("front-wheel",
                                    "rear-wheel",
                                    "4-wheel")))) +
geom_bar(position = "fill") +
scale_y_continuous(breaks = seq(0, 1, .2),
                   label = percent) +
scale_fill_brewer(palette = "Set2") +
labs(y = "Percent",
     fill="Drive Train",
     x = "Class",
     title = "Automobile Drive by Class") +
theme_minimal()
```

Each of these functions is discussed more fully in the section on Customizing graphs (see Section 11).

Next, let's add percent labels to each segment. First, we'll create a summary

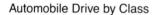

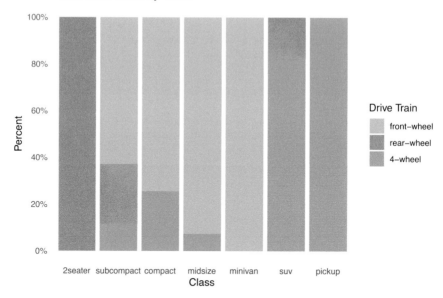

FIGURE 5.5
Segmented bar chart with improved labeling and color.

dataset that has the necessary labels.

```
# create a summary dataset
library(dplyr)
plotdata <- mpg %>%
  group_by(class, drv) %>%
  summarize(n = n()) %>%
  mutate(pct = n/sum(n),
         lbl = scales::percent(pct))
plotdata
```

```
## # A tibble: 12 x 5
## # Groups:   class [7]
##     class       drv       n     pct lbl
##     <chr>       <chr> <int>   <dbl> <chr>
## 1 2seater     r         5 1       100%
## 2 compact     4        12 0.255   26%
## 3 compact     f        35 0.745   74%
## 4 midsize     4         3 0.0732  7%
## 5 midsize     f        38 0.927   93%
## 6 minivan     f        11 1       100%
```

```
## 7 pickup     4     33 1      100%
## 8 subcompact 4      4 0.114  11%
## 9 subcompact f     22 0.629  63%
## 10 subcompact r     9 0.257  26%
## 11 suv        4     51 0.823  82%
## 12 suv        r     11 0.177  18%
```

Next, we'll use this dataset and the `geom_text` function to add labels to each bar segment.

```
# create segmented bar chart
# adding labels to each segment

ggplot(plotdata,
       aes(x = factor(class,
                      levels = c("2seater", "subcompact",
                                 "compact", "midsize",
                                 "minivan", "suv", "pickup")),
           y = pct,
           fill = factor(drv,
                         levels = c("f", "r", "4"),
                         labels = c("front-wheel",
                                    "rear-wheel",
                                    "4-wheel")))) +
  geom_bar(stat = "identity",
           position = "fill") +
  scale_y_continuous(breaks = seq(0, 1, .2),
                     label = percent) +
  geom_text(aes(label = lbl),
            size = 3,
            position = position_stack(vjust = 0.5)) +
  scale_fill_brewer(palette = "Set2") +
  labs(y = "Percent",
       fill="Drive Train",
       x = "Class",
       title = "Automobile Drive by Class") +
  theme_minimal()
```

The resulting graph (Figure 5.6) is now easy to read and interpret.

5.1.5   Other Plots

Mosaic plots provide an alternative to stacked bar charts for displaying the relationship between categorical variables. They can also provide more sophisticated statistical information. See Section 9.5 for details.

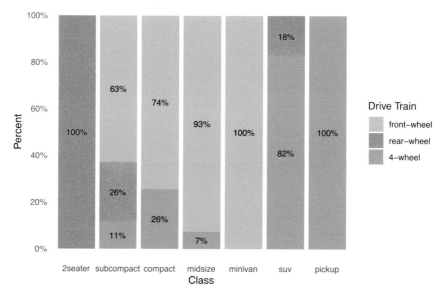

FIGURE 5.6
Segmented bar chart with value labeling.

5.2 Quantitative vs. Quantitative

The relationship between two quantitative variables is typically displayed using scatterplots and line graphs.

5.2.1 Scatterplot

The simplest display of two quantitative variables is a scatterplot, with each variable represented on an axis. Here, we will use the `Salaries` dataset described in Appendix A.1. First, let's plot experience (*yrs.since.phd*) vs. academic salary (*salary*) for college professors (Figure 5.7).

```
library(ggplot2)
data(Salaries, package="carData")

# simple scatterplot
ggplot(Salaries,
       aes(x = yrs.since.phd, y = salary)) +
  geom_point()
```

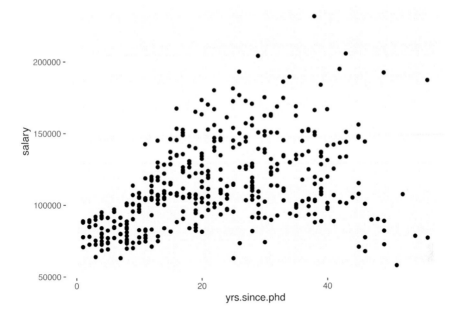

FIGURE 5.7
Simple scatterplot.

As expected, salary tends to rise with experience, but the relationship may not be strictly linear. Note that salary appears to fall off after about 40 years of experience.

The `geom_point` function has options that can be used to change:

- `color`–point color
- `size`–point size
- `shape`–point shape
- `alpha`–point transparency. Transparency ranges from 0 (transparent) to 1 (opaque), and is a useful parameter when points overlap.

The functions `scale_x_continuous` and `scale_y_continuous` control the scaling on x and y axes, respectively.

We can use these options and functions to create a more attractive scatterplot (Figure 5.8).

```
# enhanced scatter plot
ggplot(Salaries,
       aes(x = yrs.since.phd, y = salary)) +
  geom_point(color="cornflowerblue",
             size = 2,
             alpha=.8) +
```

```
scale_y_continuous(label = scales::dollar,
                   limits = c(50000, 250000)) +
scale_x_continuous(breaks = seq(0, 60, 10),
                   limits=c(0, 60)) +
labs(x = "Years Since PhD",
     y = "",
     title = "Experience vs. Salary",
     subtitle = "9-month salary for 2008-2009")
```

Again, see Customizing graphs (Chapter 11) for more details.

5.2.1.1   Adding Best Fit Lines

It is often useful to summarize the relationship displayed in the scatterplot, using a best fit line. Many types of lines are supported, including linear, polynomial, and nonparametric (loess). By default, 95% confidence limits for these lines are displayed (Figure 5.9).

```
# scatterplot with linear fit line
ggplot(Salaries, aes(x = yrs.since.phd, y = salary)) +
  geom_point(color= "steelblue") +
  geom_smooth(method = "lm")
```

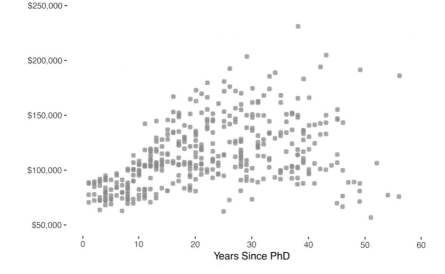

FIGURE 5.8
Scatterplot with color, transparency, and axis scaling.

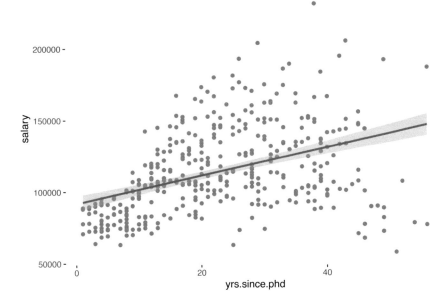

FIGURE 5.9
Scatterplot with linear fit line.

Clearly, salary increases with experience. However, there seems to be a dip at the right end–professors with significant experience, earning lower salaries. A straight line does not capture this non-linear effect. A line with a bend will fit better here.

A polynomial regression line provides a fit line of the form

$$\hat{y} = \beta_0 + \beta_1 x + \beta 2x^2 + \beta 3x^3 + \beta 4x^4 + \dots$$

Typically either a quadratic (one bend) or cubic (two bends) line is used. It is rarely necessary to use a higher order(>3) polynomials. Adding a quadratic fit line to the salary dataset produces the results in Figure 5.10.

```
# scatterplot with quadratic line of best fit
ggplot(Salaries, aes(x = yrs.since.phd, y = salary)) +
  geom_point(color= "steelblue") +
  geom_smooth(method = "lm",
              formula = y ~ poly(x, 2),
              color = "indianred3")
```

Finally, a smoothed nonparametric fit line can often provide a good picture of the relationship (Figure 5.11). The default in **ggplot2** is a loess line which stands for locally weighted scatterplot smoothing (Cleveland, 1979).

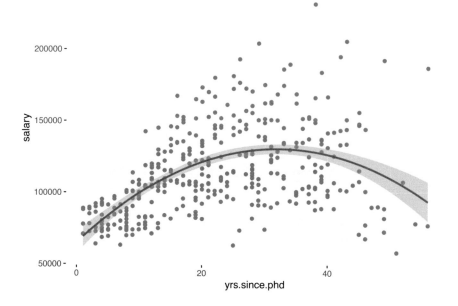

FIGURE 5.10
Scatterplot with quadratic fit line.

```
# scatterplot with loess smoothed line
ggplot(Salaries, aes(x = yrs.since.phd, y = salary)) +
  geom_point(color= "steelblue") +
  geom_smooth(color = "tomato")
```

You can suppress the confidence bands by including the option se = FALSE.
Figure 5.12 is a complete (and more attractive) plot.

```
# scatterplot with loess smoothed line
# and better labeling and color
ggplot(Salaries,
       aes(x = yrs.since.phd, y = salary)) +
  geom_point(color="cornflowerblue",
             size = 2,
             alpha=.6) +
  geom_smooth(size = 1.5,
              color = "darkgrey") +
  scale_y_continuous(label = scales::dollar,
                     limits=c(50000, 250000)) +
  scale_x_continuous(breaks = seq(0, 60, 10),
                     limits=c(0, 60)) +
```

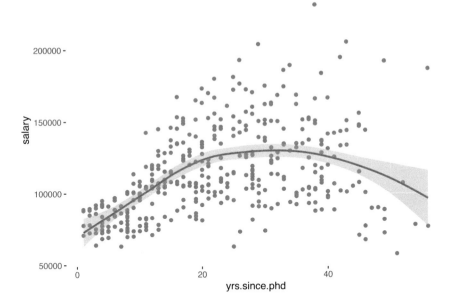

FIGURE 5.11
Scatterplot with nonparametric fit line.

```
labs(x = "Years Since PhD",
     y = "",
     title = "Experience vs. Salary",
     subtitle = "9-month salary for 2008-2009") +
theme_minimal()
```

5.2.2 Line Plot

When one of the two variables represents time, a line plot can be an effective method of displaying relationship. For example, the code below displays the relationship between time (*year*) and life expectancy (*lifeExp*) in the United States between 1952 and 2007. The data comes from the `gapminder` dataset (Appendix A.8). The graph is given in Figure 5.13.

```
data(gapminder, package="gapminder")

# Select US cases
library(dplyr)
plotdata <- filter(gapminder, country == "United States")
```

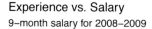

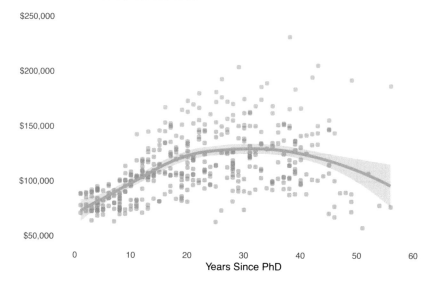

FIGURE 5.12
Scatterplot with nonparametric fit line.

```
# simple line plot
ggplot(plotdata, aes(x = year, y = lifeExp)) +
  geom_line()
```

It is hard to read indivial values in the graph above. In the next plot (Figure 5.14), we'll add points as well.

```
# line plot with points
# and improved labeling
ggplot(plotdata, aes(x = year, y = lifeExp)) +
  geom_line(size = 1.5,
            color = "lightgrey") +
  geom_point(size = 3,
             color = "steelblue") +
  labs(y = "Life Expectancy (years)",
       x = "Year",
       title = "Life expectancy changes over time",
       subtitle = "United States (1952-2007)",
       caption = "Source: http://www.gapminder.org/data/")
```

Time dependent data is covered in more detail under Time series (Section 8). Customizing line graphs is covered in the Customizing graphs (Section 11).

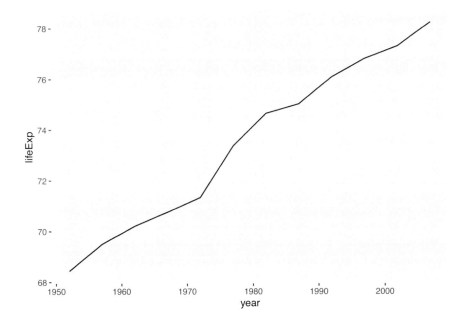

FIGURE 5.13
Simple line plot.

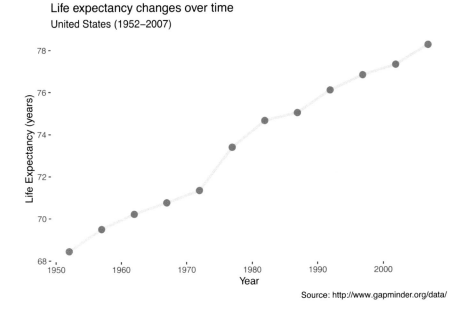

FIGURE 5.14
Line plot with points and labels.

5.3   Categorical vs. Quantitative

When plotting the relationship between a categorical variable and a quantitative variable, a large number of graph types are available. These include bar charts using summary statistics, grouped kernel density plots, side-by-side box plots, side-by-side violin plots, mean/sem plots, ridgeline plots, and Cleveland plots. Each is considered in turn.

5.3.1   Bar Chart (on Summary Statistics)

In previous sections, bar charts were used to display the number of cases by category for a single variable (Section 4.1.1) or for two variables (Section 5.1). You can also use bar charts to display other summary statistics (e.g., means or medians) on a quantitative variable for each level of a categorical variable.

For example, Figure 5.15 displays the mean salary for a sample of university professors by their academic rank.

```
data(Salaries, package="carData")

# calculate mean salary for each rank
library(dplyr)
plotdata <- Salaries %>%
  group_by(rank) %>%
  summarize(mean_salary = mean(salary))

# plot mean salaries
ggplot(plotdata, aes(x = rank, y = mean_salary)) +
  geom_bar(stat = "identity")
```

We can make it more attractive with some options. In particular, the `factor` function modifies the labels for each rank, the `scale_y_continuous` function improves the *y*-axis labeling, and the `geom_text` function adds the mean values to each bar. The results are given in Figure 5.16.

```
# plot mean salaries in a more attractive fashion
library(scales)
ggplot(plotdata,
       aes(x = factor(rank,
                      labels = c("Assistant\nProfessor",
                                 "Associate\nProfessor",
                                 "Full\nProfessor")),
           y = mean_salary)) +
  geom_bar(stat = "identity",
           fill = "cornflowerblue") +
```

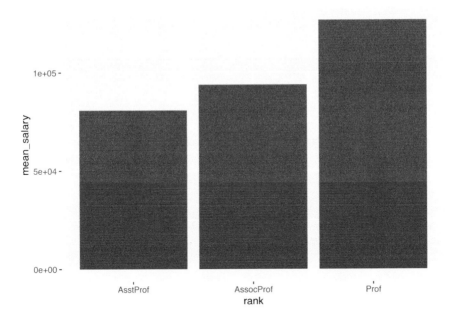

FIGURE 5.15
Bar chart displaying means.

```
geom_text(aes(label = dollar(mean_salary)),
          vjust = -0.25) +
scale_y_continuous(breaks = seq(0, 130000, 20000),
                   label = dollar) +
labs(title = "Mean Salary by Rank",
     subtitle = "9-month academic salary for 2008-2009",
     x = "",
     y = "")
```

The `vjust` parameter in the `geom_text` function controls vertical justification and nudges the text above the bars. See Annotations (Section 11.7) for more details.

One limitation of such plots is that they do not display the distribution of the data–only the summary statistic for each group. The plots below correct this limitation to some extent.

5.3.2 Grouped Kernel Density Plots

One can compare groups on a numeric variable by superimposing kernel density plots (Section 4.2.2) in a single graph (Figure 5.17).

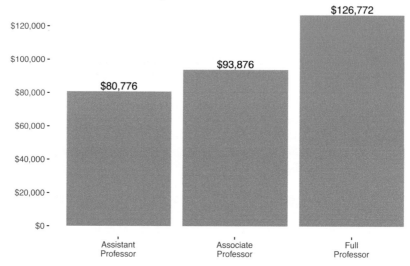

FIGURE 5.16
Bar chart displaying means.

```
# plot the distribution of salaries
# by rank using kernel density plots
ggplot(Salaries, aes(x = salary, fill = rank)) +
  geom_density(alpha = 0.4) +
  labs(title = "Salary distribution by rank")
```

The `alpha` option makes the density plots partially transparent, so that we can see what is happening under the overlaps. Alpha values range from 0 (transparent) to 1 (opaque). The graph makes clear that, in general, salary goes up with rank. However, the salary range for full professors is *very* wide.

5.3.3   Box Plots

A boxplot displays the 25th percentile, median, and 75th percentile of a distribution. The whiskers (vertical lines) capture roughly 99% of a normal distribution, and observations outside this range are plotted as points representing outliers (see the Figure 5.18).

Side-by-side box plots are very useful for comparing groups (i.e., the levels of a categorical variable) on a numerical variable. Figure 5.19 compares the distribution of salaries for the three academic ranks. The median differences, salary spreads, and outliers are clearly evident.

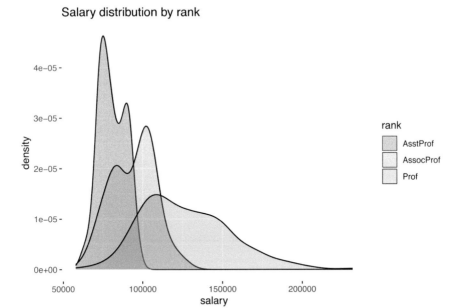

FIGURE 5.17
Grouped kernel density plots.

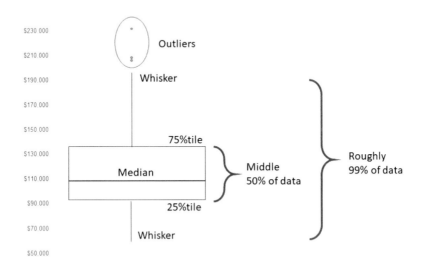

FIGURE 5.18
Side-by-side boxplots.

```
# plot the distribution of salaries by rank using boxplots
ggplot(Salaries, aes(x = rank, y = salary)) +
  geom_boxplot() +
  labs(title = "Salary distribution by rank")
```

Notched boxplots provide an approximate method for visualizing whether groups differ. Although not a formal test, if the notches of two boxplots do not overlap, there is strong evidence (95% confidence) that the medians of the two groups differ (McGill et al., 1978).

```
# plot the distribution of salaries by rank using boxplots
ggplot(Salaries, aes(x = rank, y = salary)) +
  geom_boxplot(notch = TRUE,
               fill = "cornflowerblue",
               alpha = .7) +
  labs(title = "Salary distribution by rank")
```

As shown in Figure 5.20, all three groups appear to differ.

One of the advantages of boxplots is that the width is usually not meaningful. This allows you to compare the distribution of many groups in a single graph.

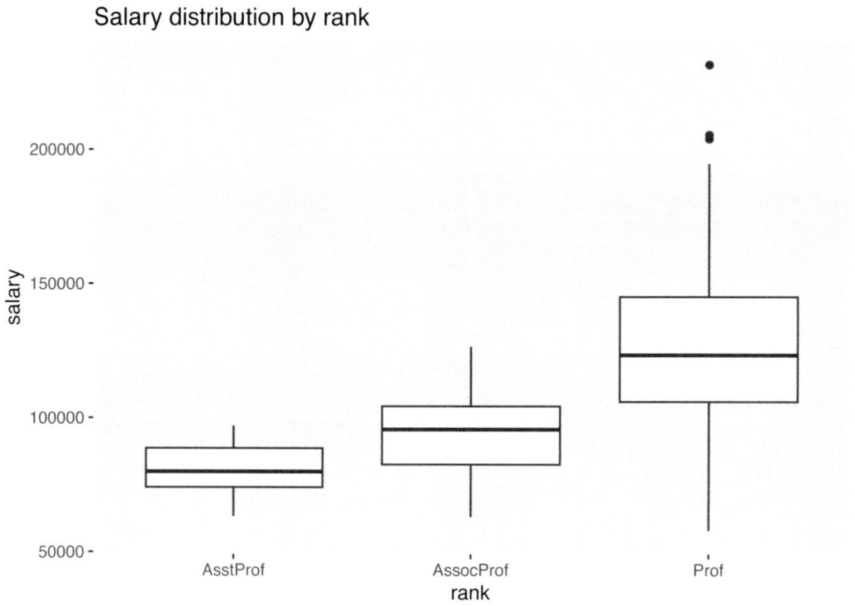

FIGURE 5.19
Side-by-side boxplots.

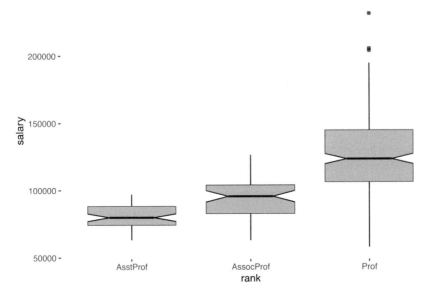

FIGURE 5.20
Side-by-side notched boxplots.

5.3.4   Violin Plots

Violin plots are similar to kernel density plots, but are mirrored and rotated 90°. An example is given in Figure 5.21.

```
# plot the distribution of salaries
# by rank using violin plots
ggplot(Salaries, aes(x = rank, y = salary)) +
  geom_violin() +
  labs(title = "Salary distribution by rank")
```

A violin plots capture more a a distribution's shape than a boxplot, but does not indicate median or middle 50% of the data. A useful variation is to superimpose boxplots on violin plots (Figure 5.22).

```
# plot the distribution using violin and boxplots
ggplot(Salaries, aes(x = rank, y = salary)) +
  geom_violin(fill = "cornflowerblue") +
  geom_boxplot(width = .15,
               fill = "orange",
               outlier.color = "orange",
```

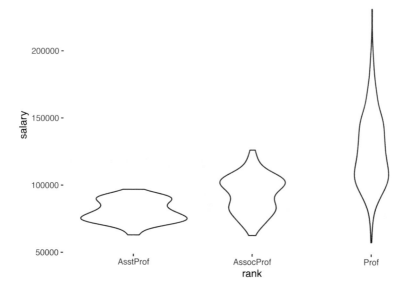

FIGURE 5.21
Side-by-side violin plots.

```
                outlier.size = 2) +
  labs(title = "Salary distribution by rank")
```

Be sure to set the `width` parameter in the `geom_boxplot` in order to assure the boxplots fit within the violin plots. You may need to play around with this in order to find a value that works well. Since geoms are layered, it is also important for the `geom_boxplot` function to appear after the `geom_violin` function. Otherwise the boxplots will be hidden beneath the violin plots.

5.3.5 Ridgeline Plots

A ridgeline plot (also called a joyplot) displays the distribution of a quantitative variable for several groups. They're similar to kernel density plots with vertical faceting, but take up less room. Ridgeline plots are created with the **ggridges** package.

Using the `mpg` dataset, let's plot the distribution of city driving miles per gallon by car class (Figure 5.23).

```
# create ridgeline graph
library(ggplot2)
library(ggridges)
```

Salary distribution by rank

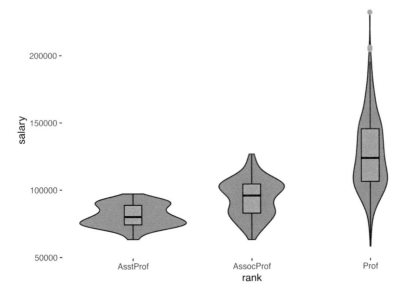

FIGURE 5.22
Side-by-side violin/box plots.

```
ggplot(mpg,
       aes(x = cty, y = class, fill = class)) +
  geom_density_ridges() +
  theme_ridges() +
  labs("Highway mileage by auto class") +
  theme(legend.position = "none")
```

I've suppressed the legend here because it's redundant (the distributions are already labeled on the y-axis). Unsurprisingly, pickup trucks have the poorest mileage, while subcompacts and compact cars tend to achieve ratings. However, there is a very wide range of gas mileage scores for these smaller cars.

Note the the possible overlap of distributions is the trade-off for a more compact graph. You can add transparency if the the overlap is severe using `geom_density_ridges(alpha = n)`, where n ranges from 0 (transparent) to 1 (opaque). See the package vignette (https://cran.r-project.org/web/packages/ggridges/vignettes/introduction.html) for more details.

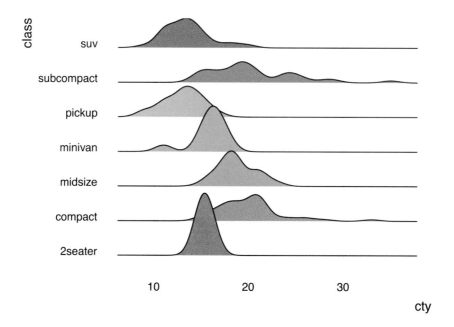

FIGURE 5.23
Ridgeline graph with color fill.

5.3.6 Mean/SEM Plots

A popular method for comparing groups on a numeric variable is a mean plot with error bars. Error bars can represent standard deviations, standard errors of the means, or confidence intervals. In this section, we'll calculate all three, but only plot means and standard errors to save space.

```
# calculate means, standard deviations,
# standard errors, and 95% confidence
# intervals by rank
library(dplyr)
plotdata <- Salaries %>%
  group_by(rank) %>%
  summarize(n = n(),
       mean = mean(salary),
       sd = sd(salary),
       se = sd / sqrt(n),
       ci = qt(0.975, df = n - 1) * sd / sqrt(n))
```

The resulting dataset is given in Table 5.1.

TABLE 5.1
Plot Data

Rank	n	Mean	sd	se	ci
AsstProf	67	80775.99	8174.113	998.6268	1993.823
AssocProf	64	93876.44	13831.700	1728.9625	3455.056
Prof	266	126772.11	27718.675	1699.5410	3346.322

```
# plot the means and standard errors
ggplot(plotdata,
       aes(x = rank,
           y = mean,
           group = 1)) +
  geom_point(size = 3) +
  geom_line() +
  geom_errorbar(aes(ymin = mean - se,
                    ymax = mean + se),
                width = .1)
```

The resulting graph appears in Figure 5.24. Although we plotted error bars representing the standard error, we could have plotted standard deviations

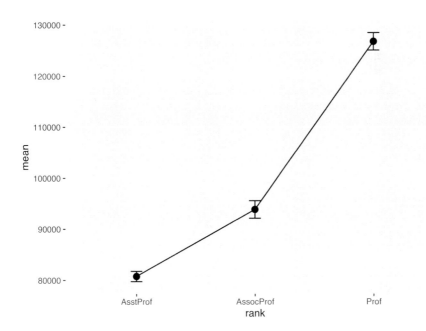

FIGURE 5.24
Mean plots with standard error bars.

or 95% confidence intervals. Simply replace se with sd or error in the aes option.

We can use the same technique to compare salary across rank and sex. (Technically, this is not bivariate since we're plotting rank, sex, and salary, but it seems to fit here.)

```
# calculate means and standard errors by rank and sex
plotdata <- Salaries %>%
  group_by(rank, sex) %>%
  summarize(n = n(),
            mean = mean(salary),
            sd = sd(salary),
            se = sd/sqrt(n))

# plot the means and standard errors by sex
ggplot(plotdata, aes(x = rank,
                     y = mean,
                     group=sex,
                     color=sex)) +
  geom_point(size = 3) +
  geom_line(size = 1) +
  geom_errorbar(aes(ymin  =mean - se,
                    ymax = mean+se),
                width = .1)
```

Unfortunately, the error bars overlap (see Figure 5.25). We can dodge the horizontal positions a bit to overcome this.

```
# plot the means and standard errors by sex (dodged)
pd <- position_dodge(0.2)
ggplot(plotdata,
       aes(x = rank,
           y = mean,
           group=sex,
           color=sex)) +
  geom_point(position = pd,
             size = 3) +
  geom_line(position = pd,
            size = 1) +
  geom_errorbar(aes(ymin = mean - se,
                    ymax = mean + se),
                width = .1,
                position= pd)
```

The dodged graph is given in Figure 5.26. Finally, lets add some options to make the graph more attractive.

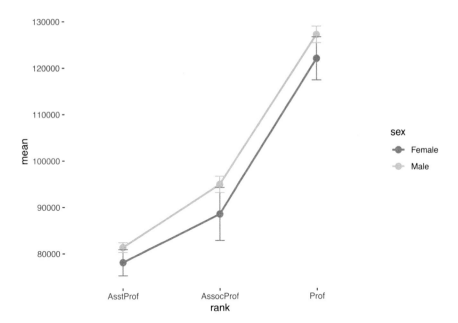

FIGURE 5.25
Mean plots with standard error bars by sex.

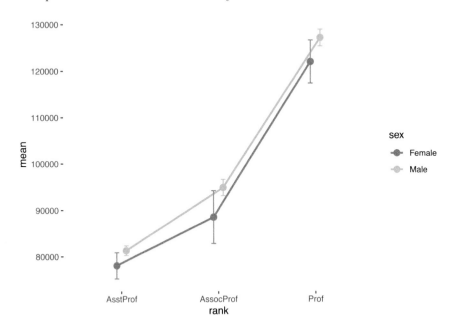

FIGURE 5.26
Mean plots with standard error bars (dodged).

```r
# improved means/standard error plot
pd <- position_dodge(0.2)
ggplot(plotdata,
       aes(x = factor(rank,
                      labels = c("Assistant\nProfessor",
                                 "Associate\nProfessor",
                                 "Full\nProfessor")),
                   y = mean, group=sex, color=sex)) +
  geom_point(position=pd,
             size=3) +
  geom_line(position=pd,
            size = 1) +
  geom_errorbar(aes(ymin = mean - se,
                    ymax = mean + se),
                width = .1,
                position=pd,
                size=1) +
  scale_y_continuous(label = scales::dollar) +
  scale_color_brewer(palette="Set1") +
  theme_minimal() +
  labs(title = "Mean salary by rank and sex",
       subtitle = "(mean +/- standard error)",
       x = "",
       y = "",
       color = "Gender")
```

Figure 5.27 is a graph you could publish in a journal.

5.3.7   Strip Plots

The relationship between a grouping variable and a numeric variable can be also displayed with a scatter plot. For example

```r
# plot the distribution of salaries
# by rank using strip plots
ggplot(Salaries, aes(y = rank, x = salary)) +
  geom_point() +
  labs(title = "Salary distribution by rank")
```

These one-dimensional scatterplots are called strip plots (Figure 5.28). Unfortunately, overprinting of points makes interpretation difficult. The relationship is easier to see if the the points are jittered (Figure 5.29). Basically, a small random number is added to each y-coordinate. To jitter the points, replace geom_point with geom_jitter.

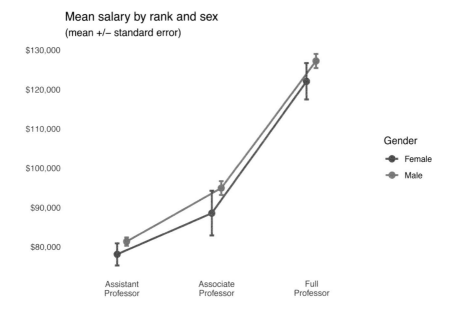

FIGURE 5.27
Mean/se plot with better labels and colors.

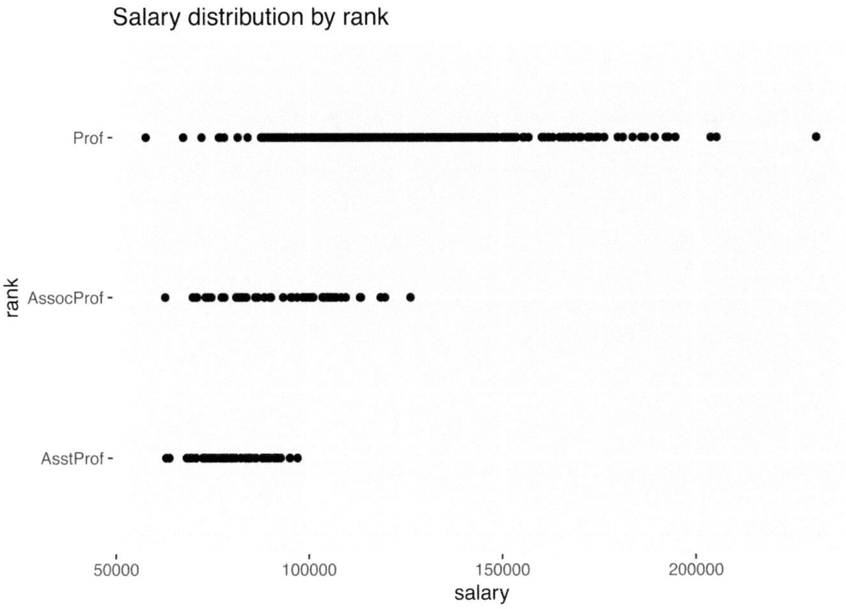

Salary distribution by rank

FIGURE 5.28
Categorical by quantiative scatterplot.

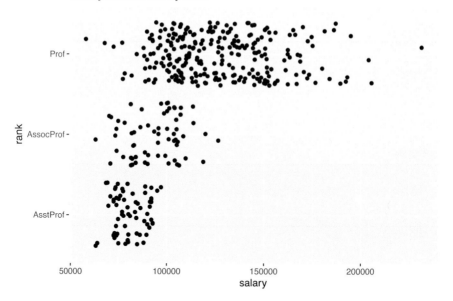

FIGURE 5.29
Jittered plot.

```r
# plot the distribution of salaries
# by rank using jittering
ggplot(Salaries, aes(y = rank, x = salary)) +
  geom_jitter() +
  labs(title = "Salary distribution by rank")
```

It is easier to compare groups if we use color (Figure 5.30).

```r
# plot the distribution of salaries
# by rank using jittering
library(scales)
ggplot(Salaries,
       aes(y = factor(rank,
                      labels = c("Assistant\nProfessor",
                                 "Associate\nProfessor",
                                 "Full\nProfessor")),
           x = salary, color = rank)) +
  geom_jitter(alpha = 0.7) +
  scale_x_continuous(label = dollar) +
  labs(title = "Academic Salary by Rank",
       subtitle = "9-month salary for 2008-2009",
```

FIGURE 5.30
Fancy jittered plot.

```
        x = "",
        y = "") +
  theme_minimal() +
  theme(legend.position = "none")
```

The option `legend.position = "none"` is used to suppress the legend (which is not needed here). Jittered plots work well when the number of points are not overly large. Here, we can not only compare groups, but see the salaries of each individual faculty member. As a college professor myself, I want to know who is making more than $200,000 on a 9-month contract!

Finally, we can superimpose boxplots on the jitter plots (Figure 5.31).

```
# plot the distribution of salaries
# by rank using jittering
library(scales)
ggplot(Salaries,
       aes(x = factor(rank,
                      labels = c("Assistant\nProfessor",
                                 "Associate\nProfessor",
                                 "Full\nProfessor")),
           y = salary, color = rank)) +
```

FIGURE 5.31
Jitter plot with superimposed box plots.

```
geom_boxplot(size=1,
             outlier.shape = 1,
             outlier.color = "black",
             outlier.size  = 3) +
geom_jitter(alpha = 0.5,
             width=.2) +
scale_y_continuous(label = dollar) +
labs(title = "Academic Salary by Rank",
     subtitle = "9-month salary for 2008-2009",
     x = "",
     y = "") +
theme_minimal() +
theme(legend.position = "none") +
coord_flip()
```

   Several options were added to create this plot.
   For the boxplot

- `size = 1` makes the lines thicker

- `outlier.color = "black"` makes outliers black
- `outlier.shape = 1` specifies circles for outliers

- `outlier.size = 3` increases the size of the outlier symbol

  For the jitter

- `alpha = 0.5` makes the points more transparent
- `width = .2` decreases the amount of jitter (.4 is the default)

Finally, the x and y axes are revered using the `coord_flip` function (i.e., the graph is turned on its side).

Before moving on, it is worth mentioning the `geom_boxjitter` function provided in the **ggpol** package. It creates a hybrid boxplot–half boxplot, half scaterplot (Figure 5.32).

```
# plot the distribution of salaries
# by rank using jittering
library(ggpol)
library(scales)
ggplot(Salaries,
       aes(x = factor(rank,
                      labels = c("Assistant\nProfessor",
                                 "Associate\nProfessor",
                                 "Full\nProfessor")),
           y = salary,
           fill=rank)) +
  geom_boxjitter(color="black",
                 jitter.color = "darkgrey",
                 errorbar.draw = TRUE) +
  scale_y_continuous(label = dollar) +
  labs(title = "Academic Salary by Rank",
       subtitle = "9-month salary for 2008-2009",
       x = "",
       y = "") +
  theme_minimal() +
  theme(legend.position = "none")
```

Choose the approach that you find most useful.

5.3.8 Cleveland Dot Charts

Cleveland plots are useful when you want to compare each observation on a numeric variable, or compare a large number of groups on a numeric summary statistic. For example, say that you want to compare the 2007 life expectancy for Asian country using the `gapminder` dataset. The Cleveland plot appears in Figure 5.33.

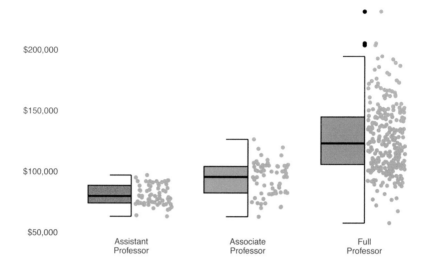

FIGURE 5.32
Using geom_boxjitter.

```
data(gapminder, package="gapminder")

# subset Asian countries in 2007
library(dplyr)
plotdata <- gapminder %>%
  filter(continent == "Asia" &
           year == 2007)

# basic Cleveland plot of life expectancy by country
ggplot(plotdata,
       aes(x= lifeExp, y = country)) +
  geom_point()
```

Comparisons are usually easier if the *y*-axis is sorted (Figure 5.34).

```
# Sorted Cleveland plot
ggplot(plotdata, aes(x=lifeExp,
                     y=reorder(country, lifeExp))) +
  geom_point()
```

The difference in life expectancy between countries like Japan and Afghanistan is striking.

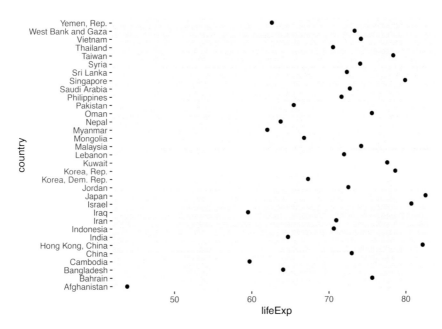

FIGURE 5.33
Basic Cleveland dot plot.

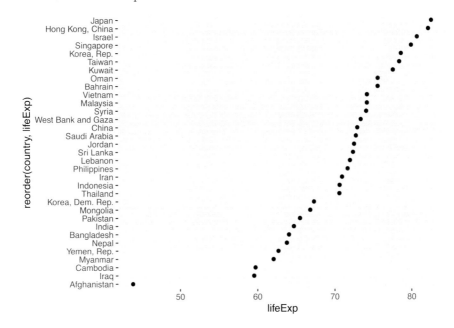

FIGURE 5.34
Sorted Cleveland dot plot.

Finally, we can use options to make the graph more attractive by removing unnecessary elements, like the grey background panel and horizontal reference lines, and adding a line segment connecting each point to the *y*-axis.

```
# Fancy Cleveland plot
ggplot(plotdata, aes(x=lifeExp,
                     y=reorder(country, lifeExp))) +
  geom_point(color="blue", size = 2) +
  geom_segment(aes(x = 40,
               xend = lifeExp,
               y = reorder(country, lifeExp),
               yend = reorder(country, lifeExp)),
               color = "lightgrey") +
  labs (x = "Life Expectancy (years)",
        y = "",
        title = "Life Expectancy by Country",
        subtitle = "GapMinder data for Asia - 2007") +
  theme_minimal() +
  theme(panel.grid.major = element_blank(),
        panel.grid.minor = element_blank())
```

This last plot is also called a lollipop graph. (I wonder why?)

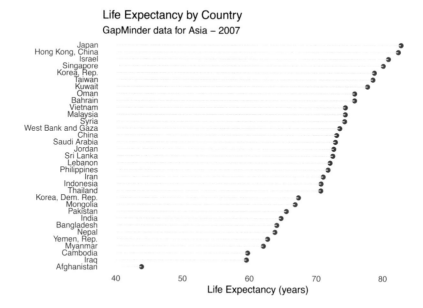

FIGURE 5.35
Fancy Cleveland plot.

6

Multivariate Graphs

In the last two chapters, you looked at ways to display the distribution of a single variable, or the relationship between two variables. We are usually interested in understanding the relations among several variables. Multivariate graphs display the relationships among three or more variables. There are two common methods for accommodating multiple variables: grouping and faceting.

6.1 Grouping

In *grouping*, the values of the first two variables are mapped to the x and y axes. Then additional variables are mapped to other visual characteristics such as color, shape, size, line type, and transparency. Grouping allows you to plot the data for multiple groups in a single graph.

Using the `Salaries` dataset, let's display the relationship between *yrs.since.phd* and *salary*. The graph is displayed in Figure 6.1.

```
library(ggplot2)
data(Salaries, package="carData")

# plot experience vs. salary
ggplot(Salaries,
       aes(x = yrs.since.phd, y = salary)) +
  geom_point() +
  labs(title = "Academic salary by years since degree")
```

Next, let's include the rank of the professor, using color (Figure 6.2).

```
# plot experience vs. salary (color represents rank)
ggplot(Salaries, aes(x = yrs.since.phd,
                     y = salary,
                     color=rank)) +
  geom_point() +
  labs(title = "Academic salary by rank and years since degree")
```

DOI: 10.1201/9781003299271-6   89

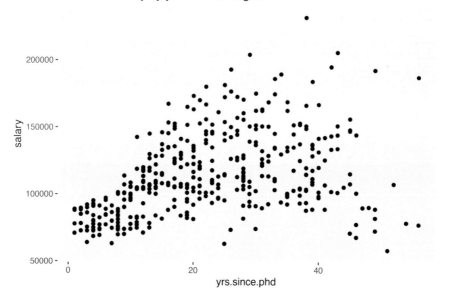

FIGURE 6.1
Simple scatterplot.

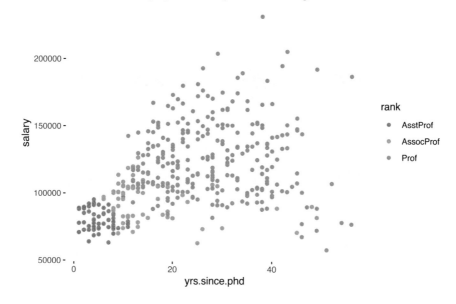

FIGURE 6.2
Scatterplot with color mapping.

Finally, let's add the gender of professor, using shape of the points to indicate sex. We'll increase the point size and transparency to make the individual points clearer. The results are given in Figure 6.3.

```
# plot experience vs. salary
# (color represents rank, shape represents sex)
ggplot(Salaries, aes(x = yrs.since.phd,
                     y = salary,
                     color = rank,
                     shape = sex)) +
  geom_point(size = 3, alpha = .6) +
  labs(title = "Academic salary by rank, sex, and years since degree")
```

Note the difference between specifying a constant value (such as `size = 3`) and a mapping of a variable to a visual characteristic (e.g., `color = rank`). Mappings are always placed within the `aes` function, while the assignment of a constant value always appear outside of the `aes` function.

Here is another example. We'll graph the relationship between years since Ph.D. and salary using the size of the points to indicate years of service. This is called a bubble plot (Figure 6.4).

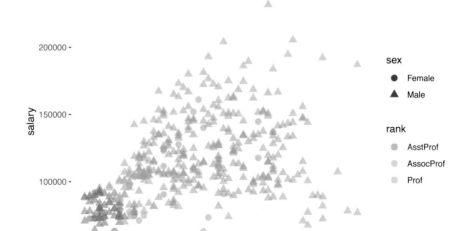

FIGURE 6.3
Scatterplot with color and shape mapping.

```
library(ggplot2)
data(Salaries, package="carData")

# plot experience vs. salary
# (color represents rank and size represents service)
ggplot(Salaries, aes(x = yrs.since.phd,
                     y = salary,
                     color = rank,
                     size = yrs.service)) +
  geom_point(alpha = .6) +
  labs(title = paste0("Academic salary by rank, years of service, ",
                      "and years since degree"))
```

Bubble plots are described in more detail in a Section 10.2.

As a final example, let's look at the *yrs.since.phd* vs *salary* and add *sex* using color and quadratic best fit lines (Figure 6.5).

```
# plot experience vs. salary with
# fit lines (color represents sex)
ggplot(Salaries,
       aes(x = yrs.since.phd,
           y = salary,
```

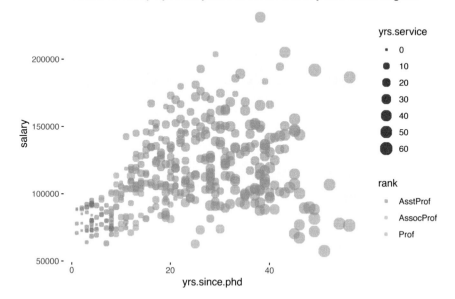

FIGURE 6.4
Scatterplot with size and color mapping.

```
            color = sex)) +
  geom_point(alpha = .4,
             size=3) +
  geom_smooth(se=FALSE,
              method="lm",
              formula=y~poly(x,2),
              size = 1.5) +
  labs(x = "Years Since Ph.D.",
       title = "Academic Salary by Sex and Years Experience",
       subtitle = "9-month salary for 2008-2009",
       y = "",
       color = "Sex") +
  scale_y_continuous(label = scales::dollar) +
  scale_color_brewer(palette="Set1") +
  theme_minimal()
```

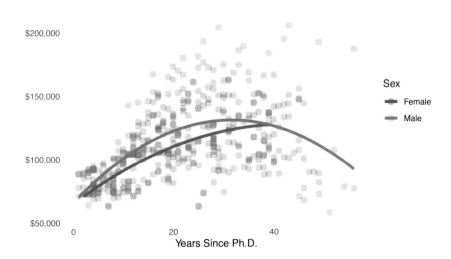

FIGURE 6.5
Scatterplot with color mapping and quadratic fit lines.

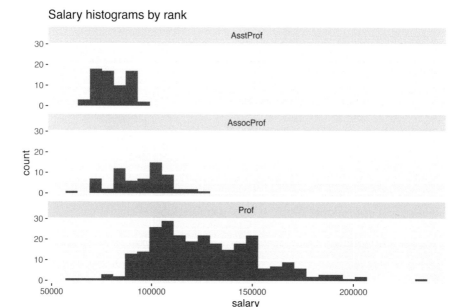

FIGURE 6.6
Salary distribution by rank.

6.2 Faceting

Grouping allows you to plot multiple variables in a single graph, using visual characteristics such as color, shape, and size. In *faceting*, a graph consists of several separate plots or *small multiples*, one for each level of a third variable, or combination of two variables. It is easiest to understand this with an example.

```
# plot salary histograms by rank
ggplot(Salaries, aes(x = salary)) +
  geom_histogram() +
  facet_wrap(~rank, ncol = 1) +
  labs(title = "Salary histograms by rank")
```

The `facet_wrap` function creates a separate graph for each level of rank (Figure 6.6). The `ncol` option controls the number of columns.

In the next example, two variables are used to define the facets.

```
# plot salary histograms by rank and sex
ggplot(Salaries, aes(x = salary/1000)) +
  geom_histogram() +
```

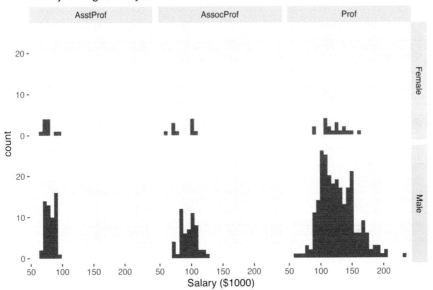

FIGURE 6.7
Salary distribution by rank and sex.

```
facet_grid(sex ~ rank) +
labs(title = "Salary histograms by sex and rank",
     x = "Salary ($1000)")
```

Here, the `facet_grid` function defines the rows (sex) and columns (rank) that separate the data into six plots in one graph (Figure 6.7).

We can also combine grouping and faceting. Figure 6.8 uses grouping (color) for sex and facets by discipline.

```
# plot salary by years of experience by sex and discipline
ggplot(Salaries,
       aes(x=yrs.since.phd, y = salary, color=sex)) +
geom_point() +
geom_smooth(method="lm",
            se=FALSE) +
facet_wrap(~discipline,
           ncol = 1)
```

Let's make this last plot more attractive. The final graph is given in Figure 6.9.

```
# plot salary by years of experience by sex and discipline
ggplot(Salaries, aes(x=yrs.since.phd,
```

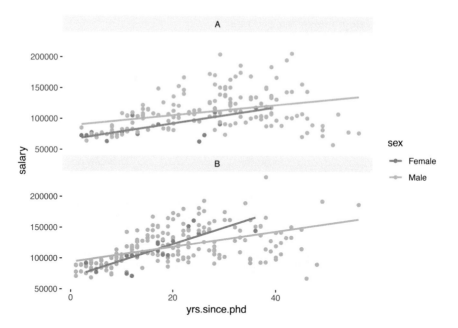

FIGURE 6.8
Salary by experience, rank, and sex.

```
                    y = salary,
                    color=sex)) +
   geom_point(size = 2,
             alpha=.5) +
   geom_smooth(method="lm",
              se=FALSE,
              size = 1.5) +
   facet_wrap(~factor(discipline,
                     labels = c("Theoretical", "Applied")),
              ncol = 1) +
   scale_y_continuous(labels = scales::dollar) +
   theme_minimal() +
   scale_color_brewer(palette="Set1") +
   labs(title = paste0("Relationship of salary and years ",
                      "since degree by sex and discipline"),
       subtitle = "9-month salary for 2008-2009",
       color = "Gender",
       x = "Years since Ph.D.",
       y = "Academic Salary")
```

See the Customizing section to learn more about customizing the appearance of a graph.

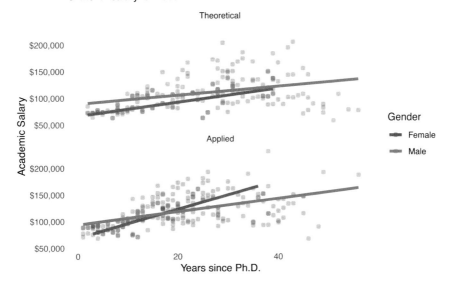

FIGURE 6.9
Salary by experience, rank, and sex (better labeled).

As a final example, we'll shift to a new dataset and plot the change in life expectancy over time for countries in the "Americas". The data comes from the gapminder dataset in the **gapminder** package. Each country appears in its own facet. The **theme** functions are used to simplify the background color, rotate the *x*-axis text, and make the font size smaller. The resulting graph is displayed in Figure 6.10.

```
# plot life expectancy by year separately
# for each country in the Americas
data(gapminder, package = "gapminder")

# Select the Americas data
plotdata <- dplyr::filter(gapminder,
                          continent == "Americas")

# plot life expectancy by year, for each country
ggplot(plotdata, aes(x=year, y = lifeExp)) +
  geom_line(color="grey") +
  geom_point(color="blue") +
  facet_wrap(~country) +
  theme_minimal(base_size = 9) +
  theme(axis.text.x = element_text(angle = 45,
```

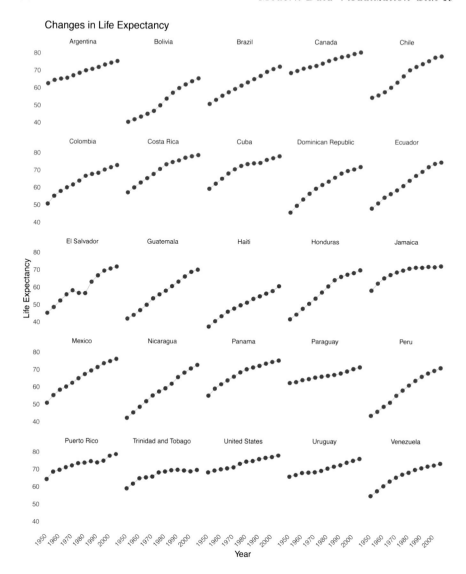

FIGURE 6.10
Changes in life expectancy by country.

```
                                        hjust = 1)) +
labs(title = "Changes in Life Expectancy",
     x = "Year",
     y = "Life Expectancy")
```

We can see that life expectancy is increasing in each country, but that Haiti is lagging behind.

Combining grouping and faceting with graphs for one (Chapter 4) or two (Chapter 5) variables allows you to create a wide range of visualizations for exploring data! You are limited only by your imagination and the over-riding goal of communicating information clearly.

7

Maps

Data are often tied to geographic locations. Examples include traffic accidents in a city, inoculation rates by state, or life expectancy by country. Viewing data superimposed onto a map can help you discover important patterns, outliers, and trends. It can also be an impactful way of conveying information to others.

In order to plot data on a map, you will need position information that ties each observation to a location. Typically position information is provided in the form of street addresses, geographic coordinates (longitude and latitude), or the names of counties, cities, or countries.

R provides a myriad of methods for creating both static and interactive maps containing spatial information. In this chapter, you'll use of **tidygeocoder**, **ggmap**, **mapview**, **choroplethr**, and **sf** to plot data onto maps.

7.1 Geocoding

Geocoding translates physical addresses (e.g., street addresses) to geographical coordinates (such as longitude and latitude.) The **tidygeocoder** package contains functions that can accomplish this translation in either direction.

Consider the following dataset.

```
location <- c("lunch", "view")
addr <- c( "10 Main Street, Middletown, CT",
          "20 W 34th St., New York, NY, 10001")
df <- data.frame(location, addr)
```

The first observation contains the street address of my favorite pizzeria. The second address is location of the Empire State Building. I can get the latitude and longitude of these addresses using the `geocode` function.

```
library(tidygeocoder)
df <- tidygeocoder::geocode(df, address = addr, method = "osm")
```

The *address* argument points to the variable containing the street address. The *method* refers to the geocoding service employed (*osm* or Open Street Maps here).

TABLE 7.1
Address Data

Location	Address
Lunch	10 Main Street, Middletown, CT
View	20 W 34th St., New York, NY, 10001

The `geocode` function supports many other services including the US Census, ArcGIS, Google, MapQuest, TomTom and others. See `?geocode` and `?geo` for details.

7.2 Dot Density Maps

Now that we know to obtain latitude/longitude from address data, let's look at dot density maps. Dot density graphs plot observations as points on a map.

The Houston crime dataset (see Appendix A.10) contains the date, time, and address of six types of criminal offenses reported between January and August 2010. We'll use this dataset to plot the locations of homicide reports.

```
library(ggmap)

# subset the data
library(dplyr)
homicide <- filter(crime, offense == "murder") %>%
  select(date, offense, address, lon, lat)

# view data
head(homicide, 3)
```

```
##       date offense          address       lon      lat
## 1 1/1/2010  murder   9650 marlive ln -95.43739 29.67790
## 2 1/5/2010  murder  1350 greens pkwy -95.43944 29.94292
## 3 1/5/2010  murder 10250 bissonnet st -95.55906 29.67480
```

TABLE 7.2
Address Data with Latitude and Longitude

Location	Address	Latitude	Longitude
Lunch	10 Main Street, Middletown, CT	41.56976	-72.64446
View	20 W 34th St., New York, NY, 10001	40.74865	-73.98530

FIGURE 7.1
Interactive map (homicide locations–default maptype).

We can create a dot density maps using either the **mapview** or **ggmap** packages. The mapview package uses the **sf** and **leaflet** packages (primarily) to quickly create interactive maps. The **ggmap** package uses **ggplot2** to creates static maps.

7.2.1   Interactive Maps with Mapview

Let's create an interactive map using the **mapview** and **sf** packages. If you are reading a hardcopy version of this chapter, be sure to run the code in order to interact with the graph. Alternatively, the online version of this book (http://rkabacoff.github.io/datavis) offers an interactive version of each graph.

First, the sf function `st_as_sf`converts the data frame to an *sf* object. An sf or *simple features* object, is a data frame containing attributes and spatial geometries that follows a widely accepted format for geographic vector data. The argument `crs = 4326` specifies a popular coordinate reference system. The `mapview` function takes this sf object and generates an interactive graph (Figure 7.1).

```
library(mapview)
library(sf)
mymap <- st_as_sf(homicide, coords = c("lon", "lat"), crs = 4326)
mapview(mymap)
```

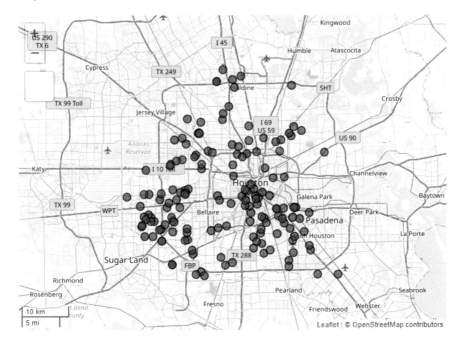

FIGURE 7.2
Interactive map (homicide locations–alternate maptype).

Clicking on a point, opens a pop-up box containing the observation's data. You can zoom in or out of the graph using the scroll wheel on your mouse, or via the + and - in the upper left corner of the plot. Below that is an option for choosing the base graph type. There is a home button in the lower right corner of the graph that resets the orientation of the graph.

There are numerous options for changing the plot. For example, let's change the point outline to black, the point fill to red, and the transparency to 0.5 (halfway between transparent and opaque). We'll also suppress the legend and home button and set the base map source to *OpenstreetMap* (see Figure 7.2).

```
library(sf)
library(mapview)
mymap <- st_as_sf(homicide, coords = c("lon", "lat"), crs = 4326)
mapview(mymap, color="black", col.regions="red",
        alpha.regions=0.5, legend = FALSE,
        homebutton = FALSE, map.types = "OpenStreetMap" )
```

Other map types include include CartoDB.Positron, CartoDB.DarkMatter, Esri.WorldImagery, and OpenTopoMap.

7.2.1.1 Using Leaflet

Leaflet (https://leafletjs.com/) is a javascript library for interactive maps and the `leaflet` package can be used to generate leaflet graphs in R. The

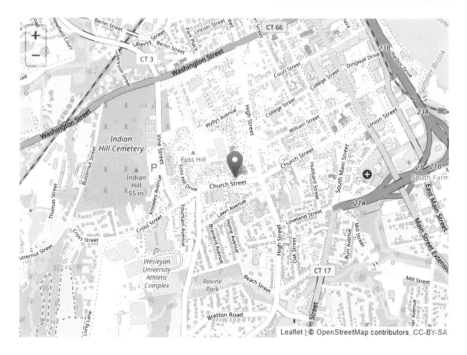

FIGURE 7.3
Interactive map (leaflet).

mapview package uses the leaflet package when creating maps. I've focused on mapview because of its ease of use. For completeness, let's use leaflet directly.

The following is a simple example. You can click on the pin, zoom in and out with the +/- buttons or mouse wheel, and drag the map around with the hand cursor. The resulting map appears in Figure 7.3.

```
# create leaflet graph
library(leaflet)
leaflet() %>%
  addTiles() %>%
  addMarkers(lng=-72.6560002,
             lat=41.5541829,
             popup="The birthplace of quantitative wisdom.</br>
             No, Waldo is not here.")
```

Leaflet allows you to create both dot density and choropleth maps. The package website (https://rstudio.github.io/leaflet/) offers a detailed tutorial and numerous examples.

7.2.2   Static Maps with ggmap

You can create a static map using the free Stamen Maps service (http://maps.stamen.com), or the paid Google maps platform (http://developers.google.com/maps). We'll consider each in turn.

7.2.2.1 Stamen Maps

As of July 31, 2023, Stamen Map Tiles are served by Stadia Maps. To create a stamen map, you will need to obtain a Stadia Maps API key. The service is free from non-commercial use.

The steps are

- Sign up for an account at stadiamaps.com.

- Go to the client dashboard. The client dashboard lets you generate, view, or revoke your API key.

- Click on "Manage Properties". Under "Authentication Configuration", generate your API key. Save this key and keep it private.

- In R, use ggmap::register_stadiamaps("your API key") to register your key.

```
ggmap::register_ stadiamaps("your API key")
```

To create a stamen map, you'll need a bounding box–the latitude and longitude for each corner of the map. The `getbb` function in the `osmdata` package can provide this.

```
# find a bounding box for Houston, Texas
library(osmdata)
bb <- getbb("houston, tx")
bb
```

```
##           min       max
## x -95.90974 -95.01205
## y  29.53707  30.11035
```

The `get_stamenmap` function takes this information and returns the map. The `ggmap` function then plots the map (Figure 7.4).

```
library(ggmap)
houston <- get_stadiamap(bbox = c(bb[1,1], bb[2,1],
                                  bb[1,2], bb[2,2]),
                         maptype="stamen_toner-lite")
ggmap(houston)
```

The map returned by the `ggmap` function is a **ggplot2** map. We can add to this graph using the standard **ggplot2** functions. In Figure 7.5, the location of homicide reports have been added.

```
# add incident locations
ggmap(houston) +
  geom_point(aes(x=lon,y=lat),data=homicide,
             color = "red", size = 2, alpha = 0.5)
```

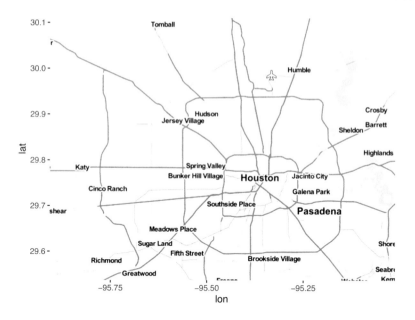

FIGURE 7.4
Static Houston map.

To clean up the results, remove the axes and add meaningful labels. The final graph is given in Figure 7.6.

```
# remove long and lat numbers and add titles
ggmap(houston) +
  geom_point(aes(x=lon,y=lat),data=homicide,
             color = "red", size = 2, alpha = 0.5) +
  theme_void() +
  labs(title = "Location of reported homocides",
       subtitle = "Houston Jan - Aug 2010",
       caption = "source: http://www.houstontx.gov/police/cs/")
```

7.2.2.2   Google Maps

To use a Google map as the base map, you will need a Google API key. Unfortunately this requires an account and valid credit card. Fortunately, Google provides a large number of uses for free, and a very reasonable rate afterward (but I take no responsibility for any costs you incur!).

Go to mapsplatform.google.com to create an account. Activate static maps and geocoding (you need to activate each separately), and receive your Google API key. Keep this API key safe and private! Once you have your key, you can create the dot density plot. The steps are listed below.

(1)   Find the center coordinates for Houston, TX.

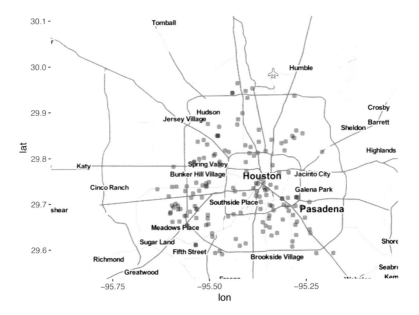

FIGURE 7.5
Houston map with crime locations.

```
library(ggmap)
# using geocode function to obtain the center coordinates
register_google(key="PutYourGoogleAPIKeyHere")
houston_center <- geocode("Houston, TX")


houston_center


##        lon        lat
## -95.36980   29.76043
```

(2) Get the background map image.

- Specify a `zoom` factor from 3 (continent) to 21 (building). The default is 10 (city).

- Specify a `maptype`. Types include terrain, terrain-background, satellite, roadmap, hybrid, watercolor, and toner.

```
# get Houston map
houston_map <- get_map(houston_center,
                       zoom = 13,
                       maptype = "roadmap")
ggmap(houston_map)
```

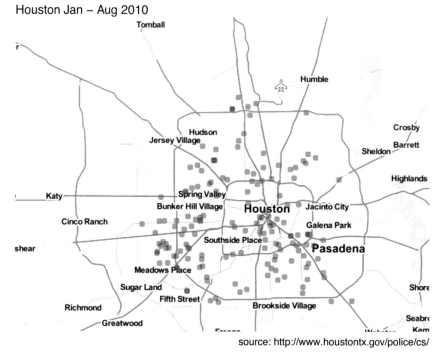

FIGURE 7.6
Crime locations with titles, and without longitude and latitude.

(3) Add crime locations to the map.

```
# add incident locations
ggmap(houston_map) +
  geom_point(aes(x=lon,y=lat),data=homicide,
             color = "red", size = 2, alpha = 0.5)
```

(4) Clean up the plot and add labels.

```
# add incident locations
ggmap(houston_map) +
  geom_point(aes(x=lon,y=lat),data=homicide,
             color = "red", size = 2, alpha = 0.5) +
  theme_void() +
  labs(title = "Location of reported homocides",
       subtitle = "Houston Jan--Aug 2010",
       caption = "source: http://www.houstontx.gov/police/cs/")
```

FIGURE 7.7
Houston map using Google Maps.

The resulting graph is given in Figure 7.7. There seems to be a concentration of homicide reports in the southern portion of the city. However, this could simply reflect population density. More investigation is needed. To learn more about ggmap, see ggmap: Spatial Visualization with ggplot2 (Kahle & Wickham, 2013).

7.3   Choropleth Maps

Choropleth maps use color or shading on predefined areas to indicate the values of a numeric variable in that area. There are numerous approaches to creating chorpleth maps. One of the easiest relies on Ari Lamstein's excellent `choroplethr` package which can create maps that display information by country, US state, and US county.

There may be times that you want to create a map for an area not covered in the choroplethr package. Additionally, you may want to create a map with greater customization. Toward this end, we'll also look at a more customizable approach using a shapefile and the **sf** and **ggplot2** packages.

FIGURE 7.8
Houston crime locations using Google Maps.

7.3.1 Data by Country

Let's create a world map and color the countries by life expectancy using the 2007 `gapminder` data.

The **choroplethr** package has numerous functions that simplify the task of creating a choropleth map. To plot the life expectancy data, we'll use the `country_choropleth` function.

The function requires that the data frame to be plotted has a column named *region* and a column named *value*. Additionally, the entries in the *region* column must exactly match how the entries are named in the *region* column of the dataset `country.map` from the **choroplethrMaps** package.

```
# view the first 12 region names in country.map
data(country.map, package = "choroplethrMaps")
head(unique(country.map$region), 12)
```

```
## [1] "afghanistan" "angola"      "azerbaijan" "moldova"   "madagascar"
## [6] "mexico"      "macedonia"   "mali"       "myanmar"   "montenegro"
## [11] "mongolia"    "mozambique"
```

Note that the region entries are all lower case.

Location of reported homocides
Houston Jan – Aug 2010

source: http://www.houstontx.gov/police/cs/

FIGURE 7.9
Customize Houston crime locations using Google Maps.

To continue, we need to make some edits to our `gapminder` dataset. Specifically, we need to

1. select the 2007 data
2. rename the *country* variable to *region*
3. rename the *lifeExp* variable to *value*
4. recode *region* values to lower cas
5. recode some *region* values to match the region values in the country.map data frame. The `recode` function in the **dplyr** package take the form recode(variable, oldvalue1 = newvalue1, oldvalue2 = newvalue2, ...)

```
# prepare dataset
data(gapminder, package = "gapminder")
plotdata <- gapminder %>%
  filter(year == 2007) %>%
  rename(region = country,
         value = lifeExp) %>%
```

```
    mutate(region = tolower(region)) %>%
    mutate(region =
      recode(region,
            "united states"     = "united states of america",
            "congo, dem. rep." = "democratic republic of the congo",
            "congo, rep."       = "republic of congo",
            "korea, dem. rep." = "south korea",
            "korea. rep."       = "north korea",
            "tanzania"          = "united republic of tanzania",
            "serbia"            = "republic of serbia",
            "slovak republic"   = "slovakia",
            "yemen, rep."       = "yemen"))
```

Now lets create the map (Figure 7.10).

```
library(choroplethr)
country_choropleth(plotdata)
```

choroplethr functions return **ggplot2** graphs. Let's make it a bit more attractive by modifying the code with additional ggplot2 functions. The resulting plot is displayed in Figure 7.11.

```
country_choropleth(plotdata,
                   num_colors=9) +
  scale_fill_brewer(palette="YlOrRd") +
  labs(title = "Life expectancy by country",
       subtitle = "Gapminder 2007 data",
       caption = "source: https://www.gapminder.org",
       fill = "Years")
```

Note that the `num_colors` option controls how many colors are used in graph. The default is seven and the maximum is nine.

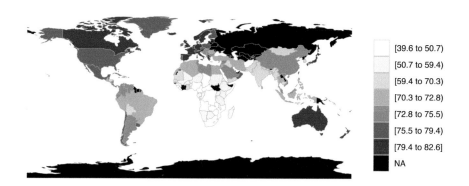

FIGURE 7.10
Choropleth map of life expectancy.

Life expectancy by country
Gapminder 2007 data

Years

[39.6 to 49.3)
[49.3 to 55.3)
[55.3 to 62.7)
[62.7 to 70.7)
[70.7 to 72.5)
[72.5 to 74.2)
[74.2 to 77.9)
[77.9 to 79.8)
[79.8 to 82.6]
NA

source: https://www.gapminder.org

FIGURE 7.11
Choropleth map of life expectancy with labels and a better color scheme.

7.3.2 Data by US State

For US data, the choroplethr package provides functions for creating maps by county, state, zip code, and census tract. Additionally, map regions can be labeled.

Let's plot US states by Hispanic and Latino populations, using the 2010 Census (see Appendix A.11). See Figure 7.12.

To plot the population data, we'll use the `state_choropleth` function. The function requires that the data frame to be plotted has a column named *region* to represent state, and a column named *value* (the quantity to be plotted). Additionally, the entries in the *region* column must exactly match how the entries are named in the *region* column of the dataset state.map from the **choroplethrMaps** package.

The `zoom = continental_us_states` option will create a map that excludes Hawaii and Alaska.

```
library(ggplot2)
library(choroplethr)
data(continental_us_states)

# input the data
library(readr)
hisplat <- read_tsv("hisplat.csv")

# prepare the data
hisplat$region <- tolower(hisplat$state)
hisplat$value <- hisplat$percent
```

```
# create the map
state_choropleth(hisplat,
                 num_colors=9,
                 zoom = continental_us_states) +
  scale_fill_brewer(palette="YlGnBu") +
  labs(title = "Hispanic and Latino Population",
       subtitle = "2010 US Census",
       caption = "source: https://tinyurl.com/2fp7c5bw",
       fill = "Percent")
```

7.3.3   Data by US County

Finally, let's plot data by US counties. We'll plot the violent crime rate per 1000 individuals for Connecticut counties in 2012. Data come from the FBI Uniform Crime Statistics. The desired graph is shown in Figure 7.13.

We'll use the `county_choropleth` function. Again, the function requires that the data frame to be plotted has a column named *region* and a column named *value*.

Additionally, the entries in the *region* column must be numeric codes and exactly match how the entries are given in the *region* column of the dataset `county.map` from the `choroplethrMaps` package.

Our dataset has country names (e.g., fairfield). However, we need region codes (e.g., 9001). We can use the `county.regions` dataset to look up the region code for each county name.

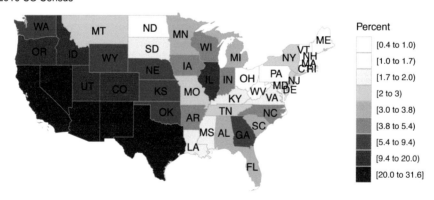

FIGURE 7.12
Choropleth map of US States.

FIGURE 7.13
County level choropleth map.

Additionally, we'll use the option `reference_map = TRUE` to add a reference map from Google Maps.

```
library(ggplot2)
library(choroplethr)
library(dplyr)

# enter violent crime rates by county
crimes_ct <- data.frame(
  county = c("fairfield", "hartford",
             "litchfield", "middlesex",
             "new haven", "new london",
             "tolland", "windham"),
  value = c(3.00, 3.32,
            1.02, 1.24,
            4.13, 4.61,
            0.16, 1.60)
)

crimes_ct
```

```
##         county value
## 1  fairfield  3.00
## 2   hartford  3.32
## 3 litchfield  1.02
## 4  middlesex  1.24
## 5  new haven  4.13
## 6 new london  4.61
## 7    tolland  0.16
## 8    windham  1.60
```

```
# obtain region codes for connecticut
data(county.regions,
     package = "choroplethrMaps")
region <- county.regions %>%
  filter(state.name == "connecticut")

region
```

```
## region county.fips.character county.name  state.name
## 1 9001             09001            fairfield connecticut
## 2 9003             09003             hartford connecticut
## 3 9005             09005           litchfield connecticut
## 4 9007             09007            middlesex connecticut
## 5 9009             09009            new haven connecticut
## 6 9011             09011           new london connecticut
## 7 9013             09013              tolland connecticut
## 8 9015             09015              windham connecticut
## state.fips.character state.abb
## 1                 09        CT
## 2                 09        CT
## 3                 09        CT
## 4                 09        CT
## 5                 09        CT
## 6                 09        CT
## 7                 09        CT
## 8                 09        CT
```

```
# join crime data to region code data
plotdata <- inner_join(crimes_ct,
                       region,
                       by=c("county" = "county.name"))
plotdata
```

```
# create choropleth map
```

```
county_choropleth(plotdata,
                  state_zoom = "connecticut",
                  reference_map = TRUE,
                  num_colors = 8) +
  scale_fill_brewer(palette="YlOrRd") +
  labs(title = "Connecticut Violent Crime Rates",
       subtitle = "FBI 2012 data",
       caption = "source: https://ucr.fbi.gov",
       fill = "Violent Crime\n Rate Per 1000")
```

See the *choroplethr help* (https://cran.r-project.org/web/packages/choroplethr/choroplethr.pdf) for more details.

7.3.4 Building a Choropleth Map Using the sf and ggplot2 Packages and a Shapefile

As stated previously, there may be times that you want to map a region not covered by the choroplethr package. Additionally, you may want greater control over the customization.

In this section, we'll create a map of the continental United States and color each states by their 2023 literacy rate (the percent of individuals who can both read and write). The literacy rates were obtained from the World Population Review (see Appendix A.7). Rather than using the choroplethr package, we'll download a US state shapefile and create the map using the **sf** and **ggplot2** packages.

(1) Prepare a shapefile

A *shapefile* is a data format that spatially describes vector features such as points, lines, and polygons. The shapefile is used to draw the geographic boundaries of the map.

You will need to find a shapefile for your the geographic area you want to plot. There are a wide range of shapefiles for cities, regions, states, and countries freely available on the internet. Natural Earth (http://naturalearthdata.com) is a good place to start. The shapefile used in the current example comes from the US Census Bureau (https://www.census.gov/geographies/mapping-files/time-series/geo/cartographic-boundary.html).

A shapefile will download as a zipped file. The code below unzips the file into a folder of the same name in the working directory (of course you can also do this by hand). The sf function `st_read` then converts the shapefile into a data frame that **ggplot2** can access.

```
library(sf)
# unzip shape file
shapefile <- "cb_2022_us_state_20m.zip"
shapedir  <- tools::file_path_sans_ext(shapefile)
```

```
if(!dir.exists(shapedir)){
  unzip(shapefile, exdir=shapedir)
}

# convert the shapefile into a data frame
# of class sf (simple features)
USMap <- st_read("cb_2022_us_state_20m/cb_2022_us_state_20m.shp")


Reading layer `cb_2022_us_state_20m' from data source
  `C:\Users\rkaba\Documents\RProjects\RKvisbook\
cb_2022_us_state_20m\cb_2022_us_state_20m.shp'
  using driver `ESRI Shapefile'
Simple feature collection with 52 features and 9 fields
Geometry type: MULTIPOLYGON
Dimension:     XY
Bounding box:  xmin: -179.1743 ymin: 17.91377
               xmax:  179.7739 ymax: 71.35256
Geodetic CRS:  NAD83


head(USMap, 3)


Simple feature collection with 3 features and 9 fields
Geometry type: MULTIPOLYGON
Dimension:     XY
Bounding box:  xmin: -124.4096 ymin: 25.84012
               xmax:  -81.9683 ymax: 42.00925
Geodetic CRS:  NAD83
  STATEFP   STATENS     AFFGEOID GEOID STUSPS
1      48 01779801 0400000US48    48     TX
2      06 01779778 0400000US06    06     CA
3      21 01779786 0400000US21    21     KY
        NAME LSAD       ALAND       AWATER
1      Texas   00 676685555821 18974391187
2 California   00 403673617862 20291712025
3   Kentucky   00 102266581101  2384240769
                        geometry
1 MULTIPOLYGON (((-106.6234 3...
2 MULTIPOLYGON (((-118.594 33...
3 MULTIPOLYGON (((-89.54443 3...
```

Note that although the sf_read function points the .shp file, all the files in the folder must be present.

The *NAME* column contains the state identifier, *STUPSPS* contains state abbreviations, and the *geometry* column is a special list object containing the coordinates need to draw the state boundaries.

(2) Prepare the data file

The literacy rates are contained in the comma delimited file named *USLitRates.csv*.

```
litRates <- read.csv("USLitRates.csv")
head(litRates, 3)
```

```
##               State Rate
## 1 New Hampshire 94.2
## 2      Minnesota 94.0
## 3  North Dakota 93.7
```

One of the most annoying aspects of creating a choropleth map is that the location variable in the data file (*State* in this case) must match the location file in the sf data frame (*NAME* in this case) exactly.

The following code will help identify any mismatches. Mismatches are printed and can be corrected.

```
# states in litRates not in USMap
setdiff(litRates$State, USMap$NAME)
```

```
## character(0)
```

We have no mismatches, so we are ready to move on.

(3) Merge the data frames

The next step combine the two data frames. Since we want to focus the on lower 48 states, we'll also eliminate Hawaii, Alaska, and Puerto Rico.

```
continentalUS <- USMap %>%
  left_join(litRates, by=c("NAME"="State")) %>%
  filter(NAME != "Hawaii" & NAME != "Alaska" &
         NAME != "Puerto Rico")
head(continentalUS, 3)
```

```
Simple feature collection with 3 features and 10 fields
Geometry type: MULTIPOLYGON
Dimension:     XY
Bounding box:  xmin: -124.4096 ymin: 25.84012
               xmax:  -81.9683 ymax: 42.00925
Geodetic CRS:  NAD83
  STATEFP  STATENS    AFFGEOID GEOID STUSPS      NAME
1      48 01779801 0400000US48    48     TX     Texas
2      06 01779778 0400000US06    06     CA California
3      21 01779786 0400000US21    21     KY   Kentucky
```

```
     LSAD          ALAND        AWATER Rate
1      00 676685555821 18974391187 81.0
2      00 403673617862 20291712025 76.9
3      00 102266581101  2384240769 87.8
                             geometry
1 MULTIPOLYGON (((-106.6234 3...
2 MULTIPOLYGON (((-118.594 33...
3 MULTIPOLYGON (((-89.54443 3...
```

(4) Create the graph (Figure 7.14)

The graph is created using **ggplot2**. Rather than specifying aes(x=, y=), aes(geometry = geometry) is used. The fill color is mapped to the literacy rate. The geom_sf function generates the map.

```
library(ggplot2)
ggplot(continentalUS, aes(geometry=geometry, fill=Rate)) +
  geom_sf()
```

(5) Customize the graph

Before finishing, lets customize the graph by

- removing the axes
- adding state labels
- modifying the fill colors and legend
- adding a title, subtitle, and caption

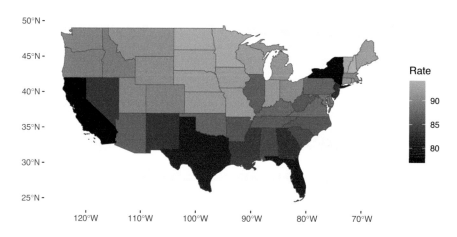

FIGURE 7.14
Choropleth map of state literacy rates.

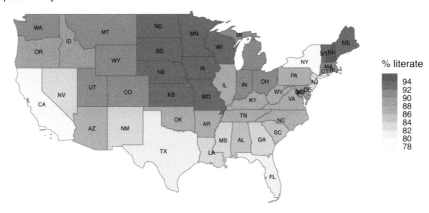

FIGURE 7.15
Customized choropleth map.

```
library(dplyr)
ggplot(continentalUS, aes(geometry=geometry, fill=Rate)) +
  geom_sf() +
  theme_void() +
  geom_sf_text(aes(label=STUSPS), size=2) +
  scale_fill_steps(low="yellow", high="royalblue",
                   n.breaks = 10) +
  labs(title="Literacy Rates by State",
       fill = "% literate",
       x = "", y = "",
       subtitle="Updated May 2023",
       caption="source: https://worldpopulationreview.com")
```

The map clearly displays the range of literacy rates among the states. Rates are lowest in New York and California.

7.4   Going Further

We've just scratched the surface of what you can do with maps in R. To learn more, see the CRAN Task View on the Analysis of Spacial Data (https://cran.r-project.org/web/views/Spatial.html) and Geocomputation with R, an comprehensive on-line book and hard-copy book (Lovelace et al., 2019).

8

Time-Dependent Graphs

A graph can be a powerful vehicle for displaying change over time. The most common time-dependent graph is the time series line graph. Other options include dumbbell charts, slope graphs, area charts, and steam graphs.

8.1 Time Series

A time series is a set of quantitative values obtained at successive time points. The intervals between time points (e.g., hours, days, weeks, months, or years) are usually equal.

Consider the Economics time series that come with the **ggplot2** package. It contains US monthly economic data collected from January 1967 to January 2015. Let's plot the personal savings rate (*psavert*) over time. We can do this with a simple line plot (Figure 8.1).

```
library(ggplot2)
ggplot(economics, aes(x = date, y = psavert)) +
  geom_line() +
  labs(title = "Personal Savings Rate",
       x = "Date",
       y = "Personal Savings Rate")
```

The `scale_x_date` function can be used to reformat dates (see Section 2.2.6). In the graph below, tick marks appear every 5 years and dates are presented in MMM-YY format. Additionally, the time series line is given an off-red color and made thicker, a nonparametric trend line (loess, Section 5.2.1.1) and titles are added, and the theme is simplified. The resulting graph is given in Figure 8.2.

```
library(ggplot2)
library(scales)
ggplot(economics, aes(x = date, y = psavert)) +
  geom_line(color = "indianred3",
            size=1 ) +
  geom_smooth() +
```

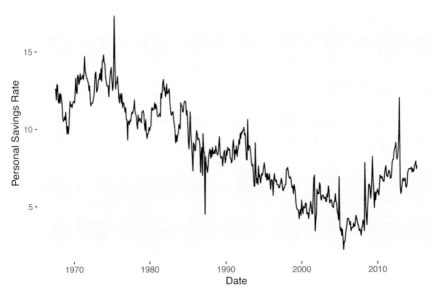

FIGURE 8.1
Simple time series.

```
scale_x_date(date_breaks = '5 years',
             labels = date_format("%b-%y")) +
labs(title = "Personal Savings Rate",
     subtitle = "1967 to 2015",
     x = "",
     y = "Personal Savings Rate") +
theme_minimal()
```

When plotting time series, be sure that the date variable is class `Date` and not class `character`. See Section 2.2.6 for details.

Let's close this section with a multivariate time series (more than one series). We'll compare closing prices for Apple and Meta from January 1, 2018 to July 31, 2023. The `getSymbols` function in the **quantmod** package is used to obtain the stock data from Yahoo Finance.

```
# multivariate time series

# one time install
# install.packages("quantmod")

library(quantmod)
```

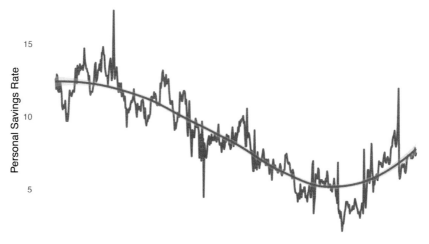

FIGURE 8.2
Simple time series with modified date axis.

```r
library(dplyr)

# get apple (AAPL) closing prices
apple <- getSymbols("AAPL",
                    return.class = "data.frame",
                    from="2023-01-01")

apple <- AAPL %>%
  mutate(Date = as.Date(row.names(.))) %>%
  select(Date, AAPL.Close) %>%
  rename(Close = AAPL.Close) %>%
  mutate(Company = "Apple")

# get Meta (META) closing prices
meta <- getSymbols("META",
                   return.class = "data.frame",
                   from="2023-01-01")

meta <- META %>%
  mutate(Date = as.Date(row.names(.))) %>%
  select(Date, META.Close) %>%
```

```
  rename(Close = META.Close) %>%
  mutate(Company = "Meta")

# combine data for both companies
mseries <- rbind(apple, meta)

# plot data
library(ggplot2)
ggplot(mseries,
       aes(x=Date, y= Close, color=Company)) +
  geom_line(size=1) +
  scale_x_date(date_breaks = '1 month',
               labels = scales::date_format("%b")) +
  scale_y_continuous(limits = c(120, 280),
                     breaks = seq(120, 280, 20),
                     labels = scales::dollar) +
  labs(title = "NASDAQ Closing Prices",
       subtitle = "Jan - June 2023",
       caption = "source: Yahoo Finance",
       y = "Closing Price") +
  theme_minimal() +
  scale_color_brewer(palette = "Dark2")
```

Figure 8.3 shows how the two stocks diverge after February. Hindsight is always 20/20!

8.2 Dumbbell Charts

Dumbbell charts are useful for displaying change between two time points for several groups or observations. The `geom_dumbbell` function from the **ggalt** package is used.

Using the **gapminder** dataset, let's plot the change in life expectancy from 1952 to 2007 in the United States. The dataset is in long format (Section 2.2.7). We will need to convert it to wide format in order to create the dumbbell plot

```
library(ggalt)
library(tidyr)
library(dplyr)

# load data
data(gapminder, package = "gapminder")
```

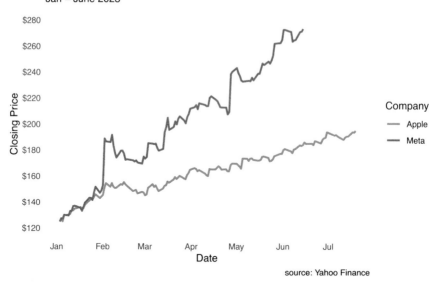

FIGURE 8.3
Multivariate time series.

```
# subset data
plotdata_long <- filter(gapminder,
                        continent == "Americas" &
                        year %in% c(1952, 2007)) %>%
  select(country, year, lifeExp)

# convert data to wide format
plotdata_wide <- pivot_wider(plotdata_long,
                             names_from = year,
                             values_from = lifeExp)
names(plotdata_wide) <- c("country", "y1952", "y2007")

# create dumbbell plot
ggplot(plotdata_wide, aes(y = country,
                          x = y1952,
                          xend = y2007)) +
  geom_dumbbell()
```

The resulting graph is given in Figure 8.3. The graph will be easier to read if the countries are sorted and the points are sized and colored. In the Figure 8.4, we'll sort by 1952 life expectancy, and modify the line and point size, color the points, add titles and labels, and simplify the theme.

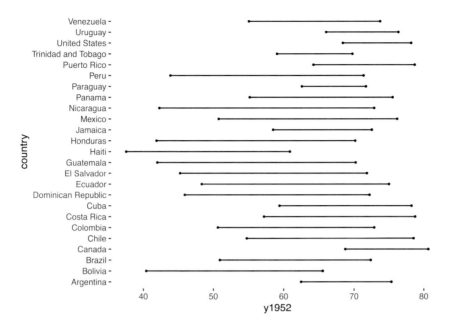

FIGURE 8.4
Simple dumbbell chart.

```
# create dumbbell plot
ggplot(plotdata_wide,
       aes(y = reorder(country, y1952),
           x = y1952,
           xend = y2007)) +
  geom_dumbbell(size = 1.2,
                size_x = 3,
                size_xend = 3,
                colour = "grey",
                colour_x = "red",
                colour_xend = "blue") +
  theme_minimal() +
  labs(title = "Change in Life Expectancy",
       subtitle = "1952 to 2007",
       x = "Life Expectancy (years)",
       y = "")
```

It is easier to discern patterns here. For example, Haiti started with the lowest life expectancy in 1952 and still has the lowest in 2007. Paraguay started relatively high by has made few gains.

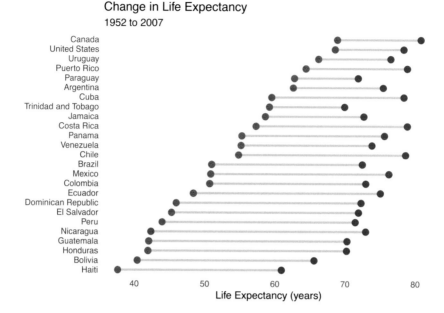

Change in Life Expectancy
1952 to 2007

FIGURE 8.5
Sorted, colored dumbbell chart.

8.3   Slope Graphs

When there are several groups and several time points, a slope graph can be
helpful. Let's plot life expectancy for six Central American countries in 1992,
1997, 2002, and 2007. Again we'll use the `gapminder` data.

To create a slope graph, we'll use the `newggslopegraph` function from the
`CGPfunctions` package.

The `newggslopegraph` function parameters are (in order)

- data frame

- time variable (which must be a factor)

- numeric variable to be plotted

- grouping variable (creating one line per group).

```
library(CGPfunctions)

# Select Central American countries data
```

```
# for 1992, 1997, 2002, and 2007

df <- gapminder %>%
  filter(year %in% c(1992, 1997, 2002, 2007) &
           country %in% c("Panama", "Costa Rica",
                          "Nicaragua", "Honduras",
                          "El Salvador", "Guatemala",
                          "Belize")) %>%
  mutate(year = factor(year),
         lifeExp = round(lifeExp))

# create slope graph

newggslopegraph(df, year, lifeExp, country) +
  labs(title="Life Expectancy by Country",
       subtitle="Central America",
       caption="source: gapminder")
```

In Figure 8.6, Costa Rica has the highest life expectancy across the range of years studied. Guatemala has the lowest, and caught up with Honduras (also low at 69) in 2002.

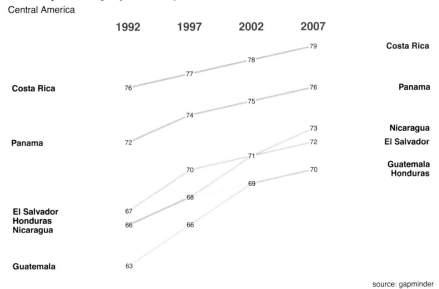

FIGURE 8.6
Slope graph.

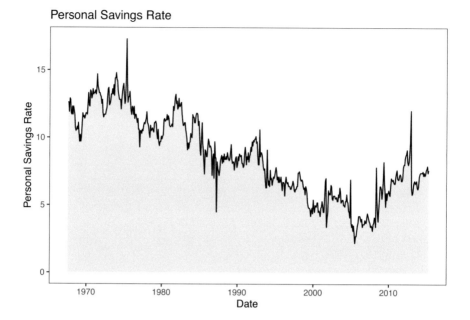

FIGURE 8.7
Basic area chart.

8.4   Area Charts

A simple area chart (Figure 8.7) is basically a line graph, with a fill from the line to the x-axis.

```
# basic area chart
ggplot(economics, aes(x = date, y = psavert)) +
  geom_area(fill="lightblue", color="black") +
  labs(title = "Personal Savings Rate",
       x = "Date",
       y = "Personal Savings Rate")
```

A stacked area chart (Figure 8.8) can be used to show differences between groups over time. Consider the uspopage dataset from the **gcookbook** package. The dataset describes the age distribution of the US population from 1900 to 2002. The variables are *year*, age group (*AgeGroup*), and number of people in thousands (*Thousands*). Let's plot the population of each age group over time.

```
# stacked area chart
data(uspopage, package = "gcookbook")
```

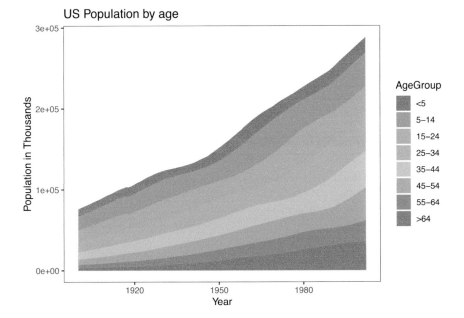

FIGURE 8.8
Stacked area chart.

```
ggplot(uspopage, aes(x = Year,
                     y = Thousands,
                     fill = AgeGroup)) +
  geom_area() +
  labs(title = "US Population by age",
       x = "Year",
       y = "Population in Thousands")
```

It is best to avoid scientific notation in your graphs. How likely is it that the average reader will know that 3e+05 means 300,000,000? It is easy to change the scale in **ggplot2**. Simply divide the thousands variable by 1000 and report it as millions. While we are at it, let's

- create black borders to highlight the difference between groups
- reverse the order the groups to match increasing age
- improve labeling
- choose a different color scheme
- choose a simpler theme.

The levels of the *AgeGroup* variable can be reversed using the `fct_rev` function in the `forcats` package.

```
# stacked area chart
data(uspopage, package = "gcookbook")
ggplot(uspopage, aes(x = Year,
                     y = Thousands/1000,
                     fill = forcats::fct_rev(AgeGroup))) +
  geom_area(color = "black") +
  labs(title = "US Population by age",
       subtitle = "1900 to 2002",
       caption = "source: U.S. Census Bureau, 2003, HS-3",
       x = "Year",
       y = "Population in Millions",
       fill = "Age Group") +
  scale_fill_brewer(palette = "Set2") +
  theme_minimal()
```

The resulting graph is given in Figure 8.9. Apparently, the number of young children have not changed very much in the past 100 years.

Stacked area charts are most useful when interest is on both (1) group change over time and (2) overall change over time. Place the most important groups at the bottom. These are the easiest to interpret in this type of plot.

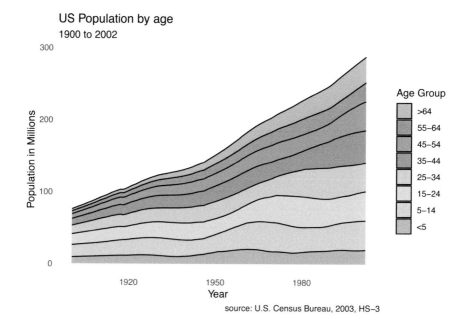

FIGURE 8.9
Stacked area chart with simpler scale.

8.5 Stream Graphs

Stream graphs (Byron & Wattenberg, 2008) are basically a variation on the stacked area chart. In a stream graph, the data is typically centered at each x-value around a mid-point and mirrored above and below that point. This is easiest to see in an example.

Let's plot the previous stacked area chart (Figure 8.9) as a stream graph (Figure 8.10).

```
# basic steam graph
data(uspopage, package = "gcookbook")
library(ggstream)
ggplot(uspopage, aes(x = Year,
                     y = Thousands/1000,
                     fill = forcats::fct_rev(AgeGroup))) +
  geom_stream() +
  labs(title = "US Population by age",
       subtitle = "1900 to 2002",
       caption = "source: U.S. Census Bureau, 2003, HS-3",
       x = "Year",
       y = "",
       fill = "Age Group") +
  scale_fill_brewer(palette = "Set2") +
  theme_minimal() +
  theme(panel.grid.major.y = element_blank(),
        panel.grid.minor.y = element_blank(),
        axis.text.y = element_blank())
```

The `theme` function is used to surpress the y-axis, whose values are not easily interpreted. To interpret this graph, look at each value on the x-axis and compare the relative vertical heights of each group. You can see, for example, that the relative proportion of older people has increased significantly.

An interesting variation is the proportional steam graph displayed in Figure 8.11. This is similar to the filled bar chart (Section 5.1.3) and makes it easier to see the relative change in values by group across time.

```
# basic stream graph
data(uspopage, package = "gcookbook")
library(ggstream)
ggplot(uspopage, aes(x = Year,
                     y = Thousands/1000,
                     fill = forcats::fct_rev(AgeGroup))) +
  geom_stream(type="proportional") +
  labs(title = "US Population by age",
```

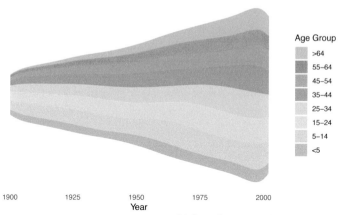

FIGURE 8.10

Basic stream graph.

```
subtitle = "1900 to 2002",
caption = "source: U.S. Census Bureau, 2003, HS-3",
x = "Year",
```

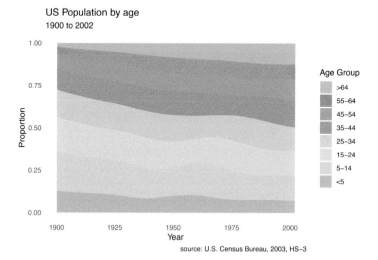

FIGURE 8.11

Proportional stream graph.

```
        y = "Proportion",
        fill = "Age Group") +
  scale_fill_brewer(palette = "Set2") +
  theme_minimal()
```

9

Statistical Models

A statistical model describes the relationship between one or more explanatory variables and one or more response variables. Graphs can help to visualize these relationships. In this section, we'll focus on models that have a single response variable that is either quantitative (a number) or binary (yes/no).

This chapter describes the use of graphs to enhance the output from statistical models. It is assumed that the reader has a passing familiarity with these models. The book R for Data Science (Wickham & Grolemund, 2017) can provide the necessary background and freely available online.

9.1   Correlation Plots

Correlation plots help you to visualize the pairwise relationships between a set of quantitative variables by displaying their correlations using color or shading.

Consider the Saratoga Houses dataset, which contains the sale price and property characteristics of Saratoga County, NY homes in 2006 (Appendix A.14). In order to explore the relationships among the quantitative variables, we can calculate the Pearson Product-Moment correlation coefficients.

In the code below, the `select_if` function in the **dplyr** package is used to select the numeric variables in the data frame. The `cor` function in base R calculates the correlations. The *use="complete.obs"* option deletes any cases with missing data. The `round` function rounds the printed results to two-decimal places.

```
data(SaratogaHouses, package="mosaicData")

# select numeric variables
df <- dplyr::select_if(SaratogaHouses, is.numeric)

# calulate the correlations
r <- cor(df, use="complete.obs")
round(r,2)
```

	price	lotSize	age	landValue	livingArea	pctCollege
price	1.00	0.16	-0.19	0.58	0.71	0.20
lotSize	0.16	1.00	-0.02	0.06	0.16	-0.03
age	-0.19	-0.02	1.00	-0.02	-0.17	-0.04
landValue	0.58	0.06	-0.02	1.00	0.42	0.23
livingArea	0.71	0.16	-0.17	0.42	1.00	0.21
pctCollege	0.20	-0.03	-0.04	0.23	0.21	1.00
bedrooms	0.40	0.11	0.03	0.20	0.66	0.16
fireplaces	0.38	0.09	-0.17	0.21	0.47	0.25
bathrooms	0.60	0.08	-0.36	0.30	0.72	0.18
rooms	0.53	0.14	-0.08	0.30	0.73	0.16

	bedrooms	fireplaces	bathrooms	rooms
price	0.40	0.38	0.60	0.53
lotSize	0.11	0.09	0.08	0.14
age	0.03	-0.17	-0.36	-0.08
landValue	0.20	0.21	0.30	0.30
livingArea	0.66	0.47	0.72	0.73
pctCollege	0.16	0.25	0.18	0.16
bedrooms	1.00	0.28	0.46	0.67
fireplaces	0.28	1.00	0.44	0.32
bathrooms	0.46	0.44	1.00	0.52
rooms	0.67	0.32	0.52	1.00

The ggcorrplot function in the **ggcorrplot** package can be used to visualize these correlations. By default, it creates a **ggplot2** graph where darker red indicates stronger positive correlations, darker blue indicates stronger negative correlations and white indicates no correlation.

```
library(ggplot2)
library(ggcorrplot)
ggcorrplot(r)
```

Figure 9.1 indicates that, an increase in number of bathrooms and living area are associated with increased price, while older homes tend to be less expensive. Older homes also tend to have fewer bathrooms.

The ggcorrplot function has a number of options for customizing the output. For example

- hc.order = TRUE reorders the variables, placing variables with similar correlation patterns together.
- type = "lower" plots the lower portion of the correlation matrix.
- lab = TRUE overlays the correlation coefficients (as text) on the plot.

```
ggcorrplot(r,
           hc.order = TRUE,
```

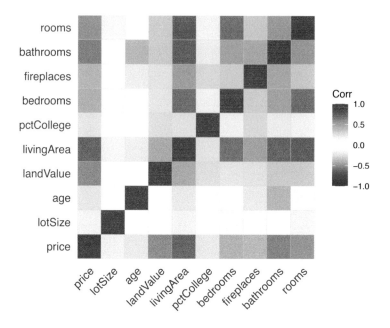

FIGURE 9.1
Correlation matrix.

```
                type = "lower",
                lab = TRUE)
```

These, and other options, produce a graph (Figure 9.2) that is easier to read and interpret. See **?ggcorrplot** for details.

9.2 Linear Regression

Linear regression allows us to explore the relationship between a quantitative response variable and an explanatory variable while other variables are held constant.

Consider the prediction of home prices in the `SaratogaHouses` dataset from lot size (square feet), age (years), land value (1000s dollars), living area (square feet), number of bedrooms and bathrooms, and whether the home is on the waterfront or not.

```
data(SaratogaHouses, package="mosaicData")
houses_lm <- lm(price ~ lotSize + age + landValue +
```

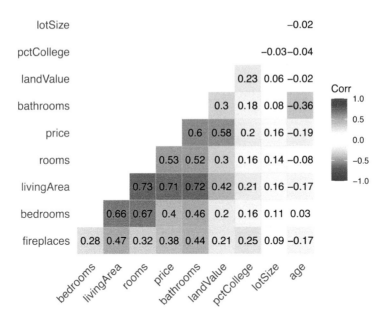

FIGURE 9.2
Sorted lower triangel correlation matrix with options.

```
                    livingArea + bedrooms + bathrooms +
                    waterfront,
                data = SaratogaHouses)
```

From the results (Table 9.1), we estimate that an increase of one square foot of living area is associated with an increase in home price of $75, holding the other variables constant. Additionally, waterfront home cost approximately $120,726 more than non-waterfront home, again controlling for the other variables in the model.

The **visreg** (http://pbreheny.github.io/visreg) package provides tools for visualizing these conditional relationships.

The `visreg` function takes (1) the model and (2) the variable of interest and plots the conditional relationship, controlling for the other variables. The option `gg = TRUE` is used to produce a **ggplot2** graph.

```
# conditional plot of price vs. living area
library(ggplot2)
library(visreg)
visreg(houses_lm, "livingArea", gg = TRUE)
```

TABLE 9.1
Linear Regression Results

Term	Estimate	std.error	Statistic	*p*.value
(Intercept)	139878.80	16472.93	8.49	0.00
lotSize	7500.79	2075.14	3.61	0.00
age	−136.04	54.16	−2.51	0.01
landValue	0.91	0.05	19.84	0.00
livingArea	75.18	4.16	18.08	0.00
bedrooms	−5766.76	2388.43	−2.41	0.02
bathrooms	24547.11	3332.27	7.37	0.00
waterfrontNo	−120726.62	15600.83	−7.74	0.00

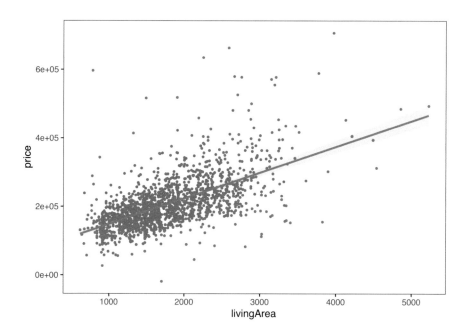

FIGURE 9.3
Conditional plot of living area and price.

Figure 9.3 suggests that, after controlling for lot size, age, living area, number of bedrooms and bathrooms, and waterfront location, sales price increases with living area in a linear fashion.

How does `visreg` work? The fitted model is used to predict values of the response variable, across the range of the chosen explanatory variable. The other variables are set to their median value (for numeric variables) or most frequent category (for categorical variables). The user can override these defaults and chose specific values for any variable in the model.

Continuing the example, the price difference between waterfront and non-waterfront homes is plotted, controlling for the other seven variables. Since a **ggplot2** graph is produced, other ggplot2 functions can be added to customize the graph.

```
# conditional plot of price vs. waterfront location
visreg(houses_lm, "waterfront", gg = TRUE) +
  scale_y_continuous(label = scales::dollar) +
  labs(title = "Relationship between price and location",
       subtitle = paste0("controlling for lot size, age, ",
                          "land value, bedrooms and bathrooms"),
       caption = "source: Saratoga Housing Data (2006)",
       y = "Home Price",
       x = "Waterfront")
```

From Figure 9.4, we can see that there are far fewer homes on the water, and they tend to be more expensive (even controlling for size, age, and land value).

The **vizreg** package provides a wide range of plotting capabilities. See *Visualization of regression models using visreg* (Breheny & Burchett, 2017) for details.

9.3 Logistic Regression

Logistic regression can be used to explore the relationship between a binary response variable and an explanatory variable while other variables are held constant. Binary response variables have two levels (yes/no, lived/died, pass/fail, malignant/benign). As with linear regression, we can use the **visreg** package to visualize these relationships.

The `CPS85` dataset in the **mosaicData** package contains a random sample of from the 1985 Current Population Survey, with data on the demographics and work experience of 534 individuals.

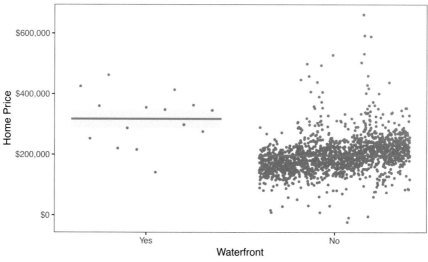

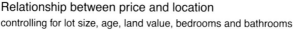

FIGURE 9.4

Conditional plot of location and price.

Let's use this data to predict the log-odds of being married, given one's sex, age, race, and job sector. We'll allow the relationship between age and marital status to vary between men and women by including an interaction term (*sex*age*).

```
# fit logistic model for predicting
# marital status: married/single
data(CPS85, package = "mosaicData")
cps85_glm <- glm(married ~ sex + age + sex*age + race + sector,
                 family="binomial",
                 data=CPS85)
```

Using the fitted model, let's visualize the relationship between age and the probability of being married, holding the other variables constant. Again, the `visreg` function takes the model and the variable of interest and plots the conditional relationship, controlling for the other variables. The option `gg = TRUE` is used to produce a **ggplot2** graph. The `scale = "response"` option creates a plot based on a probability (rather than log-odds) scale.

```
# plot results
library(ggplot2)
```

```
library(visreg)
visreg(cps85_glm, "age",
       gg = TRUE,
       scale="response") +
  labs(y = "Prob(Married)",
       x = "Age",
       title = "Relationship of age and marital status",
       subtitle = "controlling for sex, race, and job sector",
       caption = "source: Current Population Survey 1985")

## Conditions used in construction of plot
## sex: M
## race: W
## sector: prof
```

From Figure 9.5 we can see that for professional, white males, the probability of being married is roughly 0.5 at age 25 and decreases to 0.1 at age 55.

We can create multiple conditional plots by adding a by option. For example, the following code will plot the probability of being married by age, separately for men and women, controlling for race and job sector (Figure 9.6).

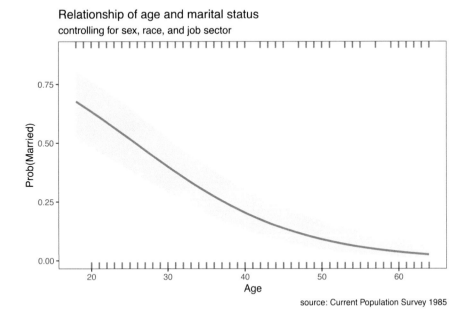

FIGURE 9.5
Conditional plot of age and marital status.

```
# plot results
library(ggplot2)
library(visreg)
visreg(cps85_glm, "age",
        by = "sex",
        gg = TRUE,
        scale="response") +
    labs(y = "Prob(Married)",
        x = "Age",
        title = "Relationship of age and marital status",
        subtitle = "controlling for race and job sector",
        caption = "source: Current Population Survey 1985")
```

In this data, the probability of marriage for men and women differ significantly over the ages measured.

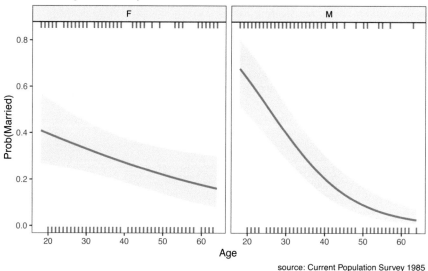

FIGURE 9.6
Conditional plot of age and marital status.

9.4   Survival Plots

In many research settings, the response variable is the time to an event. This is frequently true in healthcare research, where we are interested in time to recovery, time to death, or time to relapse.

If the event has not occurred for an observation (either because the study ended or the patient dropped out) the observation is said to be *censored.*

The NCCTG Lung Cancer dataset in the **survival** package provides data on the survival times of patients with advanced lung cancer following treatment. The study followed patients for up 34 months.

The outcome for each patient is measured by two variables:

- *time*–survival time in days

- *status* - 1 = censored, 2 = dead

Thus, a patient with *time = 305* and *status = 2* lived 305 days following treatment. Another patient with *time = 400* and *status = 1*, lived **at least** 400 days but was then lost to the study. A patient with *time = 1022* and *status = 1*, survived to the end of the study (34 months).

A survival plot (also called a Kaplan-Meier curve) can be used to illustrate the probability that an individual survives up to and including time *t*.

```
# plot survival curve
library(survival)
library(survminer)

data(lung)
sfit <- survfit(Surv(time, status) ~  1, data=lung)
ggsurvplot(sfit,
           title="Kaplan-Meier curve for lung cancer survival")
```

Figure 9.7 indicates that roughly 50% of patients are still alive 300 days post treatment. Run `summary(sfit)` for more details.

It is frequently of great interest whether groups of patients have the same survival probabilities. In the Figure 9.8, the survival curve for men and women are compared.

```
# plot survival curve for men and women
sfit <- survfit(Surv(time, status) ~  sex, data=lung)
ggsurvplot(sfit,
           conf.int=TRUE,
           pval=TRUE,
           legend.labs=c("Male", "Female"),
```

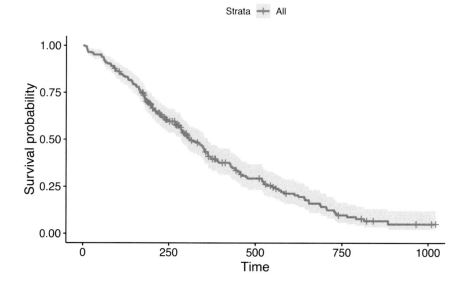

FIGURE 9.7
Basic survival curve.

```
legend.title="Sex",
palette=c("cornflowerblue", "indianred3"),
title="Kaplan-Meier Curve for lung cancer survival",
xlab = "Time (days)")
```

The `ggsurvplot` function has many options (see ?ggsurvplot). In particular, `conf.int` provides confidence intervals, while `pval` provides a log-rank test comparing the survival curves.

The p-value (0.0013) provides strong evidence that men and women have different survival probabilities following treatment. In this case, women are more likely to survive across the time period studied.

9.5   Mosaic Plots

Mosaic charts can display the relationship between categorical variables using rectangles whose areas represent the proportion of cases for any given combination of levels. The color of the tiles can also indicate the degree relationship among the variables.

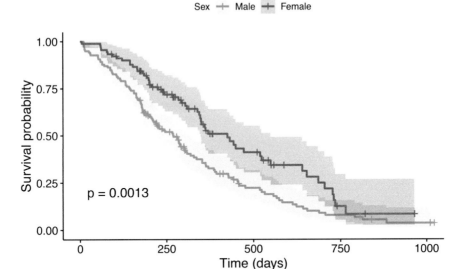

FIGURE 9.8

Comparison of survival curve.

Although mosaic charts can be created with **ggplot2** using the **ggmosaic** package, I recommend using the **vcd** package instead. Although it won't create ggplot2 graphs, the package provides a more comprehensive approach to visualizing categorical data.

People are fascinated with the Titanic (or is it with Leo?). In the Titanic disaster, what role did sex and class play in survival? We can visualize the relationship between these three categorical variables using the code below.

The dataset (`titanic.csv`) describes the sex, passenger class, and survival status for each of the 2201 passengers and crew. The `xtabs` function creates a cross-tabulation of the data, and the `ftable` function prints the results in a nice compact format.

```
# input data
library(readr)
titanic <- read_csv("titanic.csv")

# create a table
tbl <- xtabs(~Survived + Class + Sex, titanic)
ftable(tbl)

##                  Sex Female Male
## Survived Class
```

```
## No      1st           4    118
##         2nd          13    154
##         3rd         106    422
##         Crew          3    670
## Yes     1st         141     62
##         2nd          93     25
##         3rd          90     88
##         Crew         20    192
```

The `mosaic` function in the `vcd` package plots the results (Figure 9.9).

```
# create a mosaic plot from the table
library(vcd)
mosaic(tbl, main = "Titanic data")
```

The size of the tile is proportional to the percentage of cases in that combination of levels. Clearly, more passengers perished, than survived. Those that perished were primarily third class male passengers and male crew (the largest group).

If we assume that these three variables are independent, we can examine the residuals from the model and shade the tiles to match. The *shade = TRUE* adds fill colors. Dark blue represents more cases than expected given independence. Dark red represents less cases than expected if independence holds.

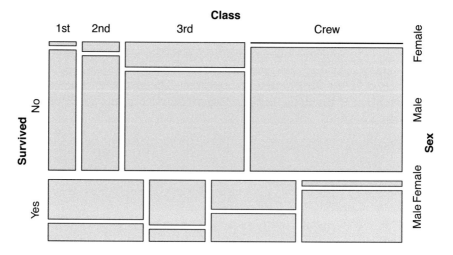

FIGURE 9.9
Basic mosaic plot.

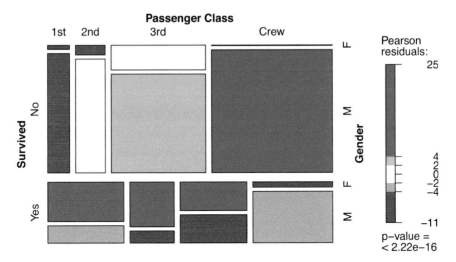

FIGURE 9.10
Mosaic plot with shading.

The *labeling_args*, `set_labels`, and `main` options improve the plot labeling (Figure 9.10).

```
mosaic(tbl,
       shade = TRUE,
       labeling_args =
         list(set_varnames = c(Sex = "Gender",
                               Survived = "Survived",
                               Class = "Passenger Class")),
       set_labels =
         list(Survived = c("No", "Yes"),
                       Class = c("1st", "2nd", "3rd", "Crew"),
                       Sex = c("F", "M")),
       main = "Titanic data")
```

We can see that if class, gender, and survival are independent, we are seeing many more male crew perishing, and first, second and third class females surviving than would be expected. Conversely, far fewer first class passengers (both male and female) died than would be expected by chance. Thus, the assumption of independence is rejected. (Spoiler alert: Leo doesn't make it.)

For complicated tables, labels can easily overlap. See `?labeling_border` for plotting options.

10

Other Graphs

Graphs in this chapter can be very useful, but don't fit in easily within the other chapters. Feel free to look through these sections and see if any of these graphs meet your needs.

10.1   3-D Scatterplot

A scatterplot displays the relationship between **two** quantitative variables (Section 5.2.1). But what do you do when you want to observe the relation between **three** variables? One approach is the 3-D scatterplot.

The **ggplot2** package and its extensions can't create a 3-D plot. However, you can create a 3-D scatterplot with the `scatterplot3d` function in the **scatterplot3d** package.

Let's say that we want to plot automobile mileage vs. engine displacement vs. car weight using the data in the mtcars data frame the comes installed with base R. The following code creates the basic 3-D scatterplot, displayed in Figure 10.1.

```
# basic 3-D scatterplot
library(scatterplot3d)
with(mtcars, {
   scatterplot3d(x = disp,
                 y = wt,
                 z = mpg,
                 main="3-D Scatterplot Example 1")
})
```

Now, lets modify the graph by replacing the points with filled blue circles, add drop lines to the $x - y$ plane, and create more meaningful labels.

```
library(scatterplot3d)
with(mtcars, {
  scatterplot3d(x = disp,
                y = wt,
```

DOI: 10.1201/9781003299271-10

3–D Scatterplot Example 1

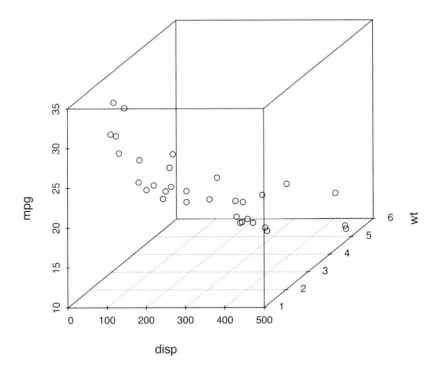

FIGURE 10.1
Basic 3-D scatterplot.

```
z = mpg,
# filled blue circles
color="blue",
pch = 19,
# lines to the horizontal plane
type = "h",
main = "3-D Scatterplot Example 2",
xlab = "Displacement (cu. in.)",
ylab = "Weight (lb/1000)",
zlab = "Miles/(US) Gallon")
})
```

In the previous code, `pch = 19` is the way we tell base R graphing function to plot points as a filled circle (pch stands for plotting character and 19 is the code for a filled circle). Similarly, `type = "h"` asks for vertical lines (like a histogram). The results are given in Figure 10.2.

3–D Scatterplot Example 2

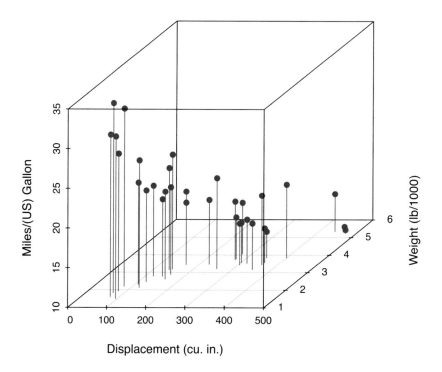

FIGURE 10.2
3-D scatterplot with vertical lines.

Next, let's label the points. We can do this by saving the results of the `scatterplot3d` function to an object, using the `xyz.convert` function to convert coordinates from 3-D (x, y, z) to 2D-projections (x, y), and apply the `text` function to add labels to the graph. The labeled graph is given in Figure 10.3.

```
library(scatterplot3d)
with(mtcars, {
  s3d <- scatterplot3d(
    x = disp,
    y = wt,
    z = mpg,
    color = "blue",
    pch = 19,
    type = "h",
    main = "3-D Scatterplot Example 3",
```

3−D Scatterplot Example 3

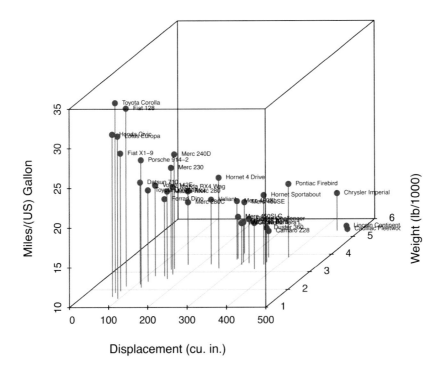

FIGURE 10.3
3-D scatterplot with vertical lines and point labels.

```
    xlab = "Displacement (cu. in.)",
    ylab = "Weight (lb/1000)",
    zlab = "Miles/(US) Gallon")

# convert 3-D coords to 2D projection
s3d.coords <- s3d$xyz.convert(disp, wt, mpg)

# plot text with 50% shrink and place to right of points
text(s3d.coords$x,
     s3d.coords$y,
     labels = row.names(mtcars),
     cex = .5,
     pos = 4)
})
```

Almost there. As a final step, we'll add information on the number of cylinders in each car. To do this, we'll add a column to the mtcars data frame indicating the color for each point. For good measure, we will shorten the *y*-axis, change the drop lines to dashed lines, and add a legend.

```
library(scatterplot3d)

# create column indicating point color
mtcars$pcolor[mtcars$cyl == 4] <- "red"
mtcars$pcolor[mtcars$cyl == 6] <- "blue"
mtcars$pcolor[mtcars$cyl == 8] <- "darkgreen"

with(mtcars, {
    s3d <- scatterplot3d(
      x = disp,
      y = wt,
      z = mpg,
      color = pcolor,
      pch = 19,
      type = "h",
      lty.hplot = 2,
      scale.y = .75,
      main = "3-D Scatterplot Example 4",
      xlab = "Displacement (cu. in.)",
      ylab = "Weight (lb/1000)",
      zlab = "Miles/(US) Gallon")

    s3d.coords <- s3d$xyz.convert(disp, wt, mpg)
    text(s3d.coords$x,
         s3d.coords$y,
         labels = row.names(mtcars),
         pos = 4,
         cex = .5)

# add the legend
legend(# top left and indented
       "topleft", inset=.05,
       # suppress legend box, shrink text 50%
       bty="n", cex=.5,
       title="Number of Cylinders",
       c("4", "6", "8"),
       fill=c("red", "blue", "darkgreen"))
})
```

We can easily see from Figure 10.4 that the car with the highest mileage (Toyota Corolla) has low engine displacement, low weight, and four cylinders.

3-D scatterplots can be difficult to interpret because they are static. The `scatter3d` function in the **car** package allows you to create an interactive 3-D

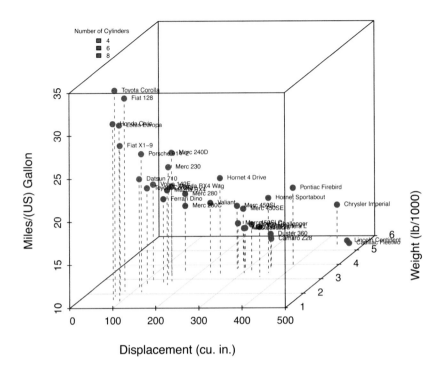

3–D Scatterplot Example 4

FIGURE 10.4

3-D scatterplot with vertical lines and point labels and legend.

graph that can be manually rotated.

```
library(car)
with(mtcars,
     scatter3d(disp, wt, mpg))
```

You can now use your mouse to rotate the axes and zoom in and out with the mouse scroll wheel (Figure 10.5). Note that this will only work if you actually run the code on your desktop. If you are trying to manipulate the graph in this book you'll drive yourself crazy!

The graph can be highly customized. In the next graph (Figure 10.6),

- each axis is colored black
- points are colored red for automatic transmission and blue for manual transmission
- all 32 data points are labeled with their rowname

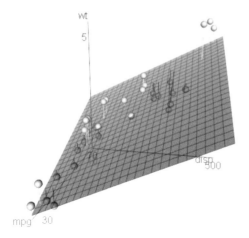

FIGURE 10.5
Interative 3-D scatterplot.

- the default best fit surface is suppressed
- all three axes are given longer labels

```
library(car)
with(mtcars,
     scatter3d(disp, wt, mpg,
               axis.col = c("black", "black", "black"),
               groups = factor(am),
               surface.col = c("red", "blue"),
               col = c("red", "blue"),
               text.col = "grey",
               id = list(n=nrow(mtcars),
                         labels=rownames(mtcars)),
               surface = FALSE,
               xlab = "Displacement",
               ylab = "Weight",
               zlab = "Miles Per Gallon"))
```

The `id` option consists of a list of options that control the identification of points. If `n` is less than the total number of points, the n most extreme points are labeled. Here, all points are labeled and the row.names are used for the labels.

The `axis.col` option specifies the color of the x, y, and z axis, respectively. The `surface.col` option specifies the colors of the points by group. The *col* option specifies by colors of the point labels by group. The `text.col` option specifies the color of the axis labels.

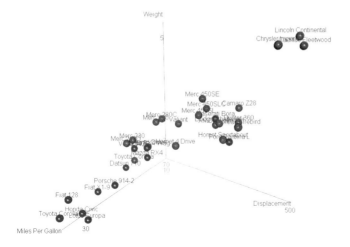

FIGURE 10.6
Interative 3-D scatterplot with labeled points.

The `surface` option indicates if a surface fit should be plotted (default = TRUE) and the `lab` options adds labels to the axes.

Labeled 3-D scatterplots are most effective when the number of labeled points is small. Otherwise label overlap becomes a significant issue.

10.2   Bubble Charts

A bubble chart is also useful when plotting the relationship between **three** quantitative variables. A bubble chart is basically just a scatterplot where the point size is proportional to the values of a third quantitative variable.

Using the mtcars dataset, let's plot car weight vs. mileage and use point size to represent horsepower. The graph is displayed in Figure 10.7.

```
# create a bubble plot
data(mtcars)
library(ggplot2)
ggplot(mtcars,
        aes(x = wt, y = mpg, size = hp)) +
  geom_point()
```

We can improve the default appearance by increasing the size of the bubbles, choosing a different point shape and color, and adding some transparency.

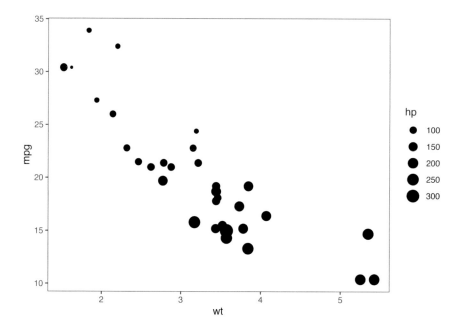

FIGURE 10.7
Basic bubble plot.

```
# create a bubble plot with modifications
ggplot(mtcars,
       aes(x = wt, y = mpg, size = hp)) +
  geom_point(alpha = .5,
             fill="cornflowerblue",
             color="black",
             shape=21) +
  scale_size_continuous(range = c(1, 14)) +
  labs(title = "Auto mileage by weight and horsepower",
       subtitle = "Motor Trend US Magazine (1973-74 models)",
       x = "Weight (1000 lbs)",
       y = "Miles/(US) gallon",
       size = "Gross horsepower")
```

The `range` parameter in the `scale_size_continuous` function specifies the minimum and maximum size of the plotting symbol. The default is `range = c(1, 6)`.

The `shape` option in the `geom_point` function specifies an circle with a border color and fill color.

From Figure 10.8 it is clear that miles per gallon decreases with increased car weight and horsepower. However, there is one car with low weight, high horsepower, and high gas mileage. Going back to the data, it's the Lotus

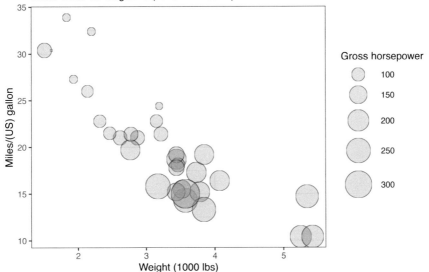

FIGURE 10.8
Bubble plot with modifications.

Europa.

Bubble charts are controversial for the same reason that pie charts are controversial. People are better at judging length than volume. However, they are quite popular.

10.3 Biplots

3-D scatterplots and bubble charts plot the relation between three quantitative variables. With more than three quantitative variables, a biplot (Nishisato et al., 2021) can be very useful. A biplot is a specialized graph that attempts to represent the relationship between observations, between variables, and between observations and variables, in a low (usually two) dimensional space.

It's easiest to see how this works with an example. Let's create a biplot for the mtcars dataset, using the fviz_pca function from the **factoextra** package.

```
# create a biplot
# load data
```

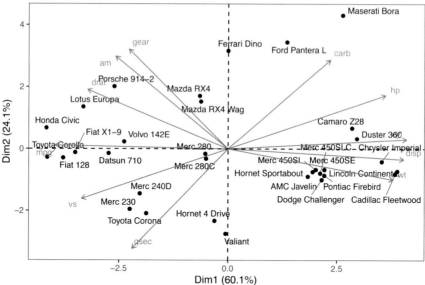

FIGURE 10.9
Basic biplot.

```
data(mtcars)

# fit a principal components model
fit <- prcomp(x = mtcars,
              center = TRUE,
              scale = TRUE)

# plot the results
library(factoextra)
fviz_pca(fit,
         repel = TRUE,
         labelsize = 3) +
  theme_bw() +
  labs(title = "Biplot of mtcars data")
```

The `fviz_pca` function produces a **ggplot2** graph (Figure 10.9).

Dim1 and *Dim2* are the first two principal components–linear combinations of the original p variables.

$$PC_1 = \beta_{10} + \beta_{11}x_1 + \beta_{12}x_2 + \beta_{13}x_3 + \cdots + \beta_{1p}x_p$$
$$PC_2 = \beta_{20} + \beta_{21}x_1 + \beta_{22}x_2 + \beta_{23}x_3 + \cdots + \beta_{2p}x_p$$

The weights of these linear combinations ($\beta_{ij}s$) are chosen to maximize the variance accounted for in the original variables. Additionally, the principal components (PCs) are constrained to be uncorrelated with each other.

In this graph, the first PC accounts for 60% of the variability in the original data. The second PC accounts for 24%. Together, they account for 84% of the variability in the original $p = 11$ variables.

As you can see, both the observations (cars) and variables (car characteristics) are plotted in the same graph.

- Points represent observations. Smaller distances between points suggest similar values on the original set of variables. For example, the *Toyota Corolla* and *Honda Civic* are similar to each other, as are the *Chrysler Imperial* and *Liconln Continental*. However, the *Toyota Corolla* is very different from the *Lincoln Continental*.
- The vectors (arrows) represent variables. The angle between vectors are proportional to the correlation between the variables. Smaller angles indicate stronger correlations. For example, *gear* and *am* are positively correlated, *gear* and *qsec* are uncorrelated (90 degree angle), and *am* and *wt* are negatively correlated (angle greater than 90 degrees).

- The observations that are are farthest along the direction of a variable's vector, have the highest values on that variable. For example, the *Toyota Corolla* and *Honda Civic* have higher values on *mpg*. The *Toyota Corona* has a higher *qsec*. The *Duster 360* has more *cylinders*.

As you can see, biplots convey an amazing amount of information in a single graph. However, care must be taken in interpreting biplots. They are only accurate when the percentage of variance accounted for is high. Always check your conclusion with the original data. For example, if the graph suggests that two cars are similar, go back to the original data and do a spot-check to see if that is so.

See the article by Forrest Young (https://www.uv.es/visualstats/vista-frames/help/lecturenotes/lecture13/biplot.html) to learn more about interpreting biplots correctly.

A flow diagram represents a set of dynamic relationships. It usually captures the physical or metaphorical flow of people, materials, communications, or objects through a set of nodes in a network.

10.4 Alluvial Diagrams

Alluvial diagrams are useful for displaying the relation among two or more categorical variables. They use a flow analogy to represent in changes in group

composition across variables. This will be more understandable when you see
an example.

Alluvial diagrams are created with **ggalluvial** package, generating **ggplot2**
graphs. As an example, let's diagram the survival of Titanic passengers, using
the `titanic` dataset. We will look at the relationship between passenger class,
sex, and survival.

To create an alluvial diagram, first count the frequency of of each combination of the categorical variables.

```
# input data
library(readr)
titanic <- read_csv("titanic.csv")

# summarize data
library(dplyr)
titanic_table <- titanic %>%
  group_by(Class, Sex, Survived) %>%
  count()

# convert survived to a factor with labels
titanic_table$Survived <- factor(titanic_table$Survived,
                                 levels = c("Yes", "No"))
# view the first 6 cases
head(titanic_table)

## # A tibble: 6 x 4
## # Groups:   Class, Sex, Survived [6]
##    Class Sex    Survived     n
##    <chr> <chr> <fct>    <int>
## 1 1st   Female No           4
## 2 1st   Female Yes        141
## 3 1st   Male   No         118
## 4 1st   Male   Yes         62
## 5 2nd   Female No          13
## 6 2nd   Female Yes         93
```

Next create an alluvial diagram in **ggplot2** using the `ggplot`,
`geom_alluvium` and `geom_stratum` functions. The categorical variables are
mapped to *axes* and n to *y*. This will produce Figure 10.10

```
library(ggalluvial)
ggplot(titanic_table,
       aes(axis1 = Class,
           axis2 = Sex,
           axis3 = Survived,
```

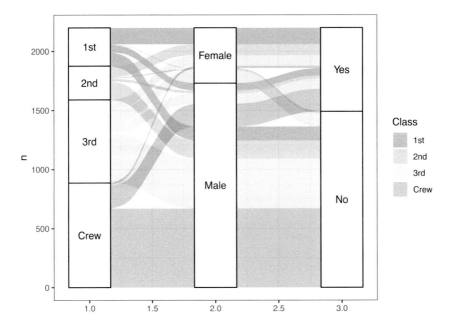

FIGURE 10.10
Basic alluvial diagram.

```
              y = n)) +
    geom_alluvium(aes(fill = Class)) +
    geom_stratum() +
    geom_text(stat = "stratum",
              aes(label = after_stat(stratum)))
```

To interpret the graph (Figure 10.10), start with the variable on the left (*Class*) and follow the flow to the right. The height of the category level represents the proportion of observations in that level. For example, the crew made up roughly 40% of the passengers. Roughly, 30% of passengers survived.

The height of the flow represents the proportion of observations contained in the two variable levels they connect. About 50% of first-class passengers were females, and all female first-class passengers survived. The crew was overwhelmingly male, and roughly 75% of this group perished.

As a second example, let's look at the relationship between the number carburetors, cylinders, gears, and the transmission type (manual or automatic) for the 32 cars in the mtcars dataset. We'll treat each variable as categorical.

First, we need to prepare the data.

```
library(dplyr)
data(mtcars)
mtcars_table <- mtcars %>%
  mutate(am = factor(am, labels = c("Auto", "Man")),
         cyl = factor(cyl),
         gear = factor(gear),
         carb = factor(carb)) %>%
  group_by(cyl, gear, carb, am) %>%
  count()

head(mtcars_table)

## # A tibble: 6 x 5
## # Groups:    cyl, gear, carb, am [6]
##    cyl   gear  carb  am        n
##    <fct> <fct> <fct> <fct> <int>
## 1 4      3     1     Auto      1
## 2 4      4     1     Man       4
## 3 4      4     2     Auto      2
## 4 4      4     2     Man       2
## 5 4      5     2     Man       2
## 6 6      3     1     Auto      2
```

Next, create the graph. Several options and functions are added to enhance the results. Specifically,

- the flow borders are set to black (`geom_alluvium`)
- the strata are given transparency (`geom_strata`)
- the strata are labeled and made wider (`scale_x_discrete`)
- titles are added (`labs`)
- the theme is simplified (`theme_minimal`)
- the legend is suppressed (`theme`)

```
ggplot(mtcars_table,
       aes(axis1 = carb,
           axis2 = cyl,
           axis3 = gear,
           axis4 = am,
           y = n)) +
  geom_alluvium(aes(fill = carb), color="black") +
  geom_stratum(alpha=.8) +
  geom_text(stat = "stratum",
            aes(label = after_stat(stratum))) +
  scale_x_discrete(limits = c("Carburetors", "Cylinders",
                              "Gears", "Transmission"),
```

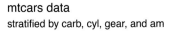

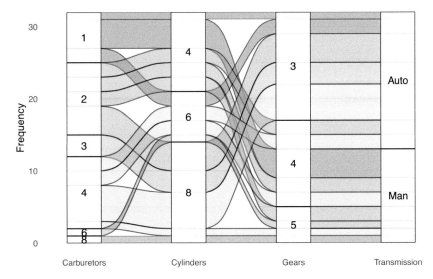

FIGURE 10.11
Customized alluvial diagram.

```
                        expand = c(.1, .1)) +
    # scale_fill_brewer(palette="Paired") +
    labs(title = "mtcars data",
          subtitle = "stratified by carb, cyl, gear, and am",
          y = "Frequency") +
    theme_minimal() +
    theme(legend.position = "none")
```

I think that these changes make the graph (Figure 10.11) easier to follow. For example, all eight carburetor cars have eight cylinders, five gears, and one manual transmission. Most four carburetor cars have eight cylinders, three gears, and an automatic transmission.

See the ggalluvial website (https://github.com/corybrunson/ggalluvial) for additional details.

10.5 Heatmaps

A heatmap displays a set of data using colored tiles for each variable value within each observation. There are many varieties of heatmaps. Although base

R comes with a `heatmap` function, we'll use the more powerful **superheat** package (I love these names).

First, let's create a heatmap for the mtcars dataset that come with base R. The mtcars dataset contains information on 32 cars measured on 11 variables.

```
# create a heatmap
data(mtcars)
library(superheat)
superheat(mtcars, scale = TRUE)
```

The `scale = TRUE` options standardizes the columns to a mean of zero and standard deviation of one. Looking at the graph (Figure 10.12), we can see that the Merc 230 has a quarter mile time (*qsec*) the is well above average (bright yellow). The Lotus Europa has a weight is well below average (dark blue).

We can use clustering to sort the rows and/or columns. In the next example (Figure 10.13), we'll sort the rows so that cars that are similar appear near each other. We will also adjust the text and label sizes.

```
# sorted heat map
superheat(mtcars,
          scale = TRUE,
          left.label.text.size=3,
          bottom.label.text.size=3,
          bottom.label.size = .05,
          row.dendrogram = TRUE )
```

Here we can see that the Toyota Corolla and Fiat 128 have similar characteristics. The Lincoln Continental and Cadillac Fleetwood also have similar characteristics.

The **superheat** function requires that the data be in particular format. Specifically,

- the data must be all numeric

- the row names are used to label the left axis. If the desired labels are in a column variable, the variable must be converted to row names (more on this below)

- missing values are allowed

Let's use a heatmap to display changes in life expectancies over time for Asian countries. The data come from the `gapminder` dataset (Appendix A.8).

Since the data is in long format (Section 2.2.7), we first have to convert to wide format. Then we need to ensure that it is a data frame and convert the variable *country* into row names. Finally, we'll sort the data by 2007 life expectancy. While we are at it, let's change the color scheme.

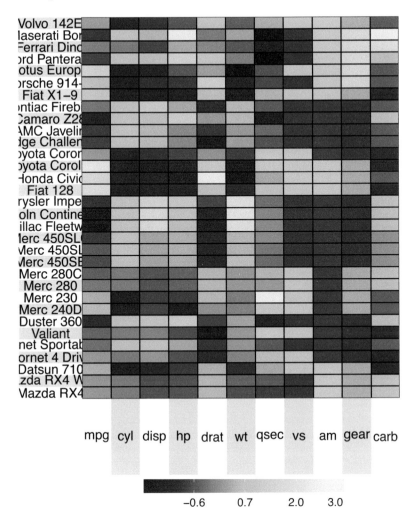

FIGURE 10.12
Basic heatmap.

```
# create heatmap for gapminder data (Asia)
library(tidyr)
library(dplyr)

# load data
data(gapminder, package="gapminder")

# subset Asian countries
```

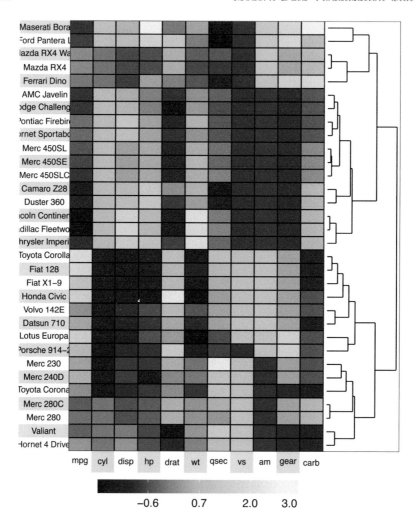

FIGURE 10.13
Sorted heatmap.

```
asia <- gapminder %>%
  filter(continent == "Asia") %>%
  select(year, country, lifeExp)

# convert to long to wide format
plotdata <- pivot_wider(asia, names_from = year,
                        values_from = lifeExp)

# save country as row names
```

```
plotdata <- as.data.frame(plotdata)
row.names(plotdata) <- plotdata$country
plotdata$country <- NULL

# row order
sort.order <- order(plotdata$"2007")

# color scheme
library(RColorBrewer)
colors <- rev(brewer.pal(5, "Blues"))


# create the heat map
superheat(plotdata,
          scale = FALSE,
          left.label.text.size=3,
          bottom.label.text.size=3,
          bottom.label.size = .05,
          heat.pal = colors,
          order.rows = sort.order,
          title = "Life Expectancy in Asia")
```

The resulting graph is given in Figure 10.14. Japan, Hong Kong, and Israel have the highest life expectancies. South Korea was doing well in the 80s but has lost some ground. Life expectancy in Cambodia took a sharp hit in 1977.

To see what you can do with heat maps, see the extensive **superheat** vignette (https://rlbarter.github.io/superheat/).

10.6   Radar Charts

A radar chart (also called a spider or star chart) displays one or more groups or observations on three or more quantitative variables.

In the example below, we'll compare dogs, pigs, and cows in terms of body size, brain size, and sleep characteristics (total sleep time, length of sleep cycle, and amount of REM sleep). The data come from the `msleep` dataset that ships with **ggplot2**.

Radar charts can be created with `ggradar` function in the **ggradar** package. Next, we have to put the data in a specific format:

- The first variable should be called *group* and contain the identifier for each observation

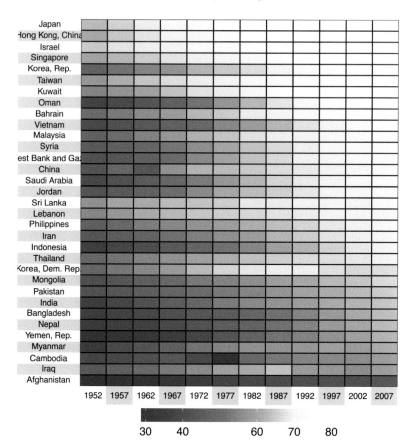

FIGURE 10.14
Heatmap for time series.

- The numeric variables have to be rescaled so that their values range from 0 to 1

```
# create a radar chart

# prepare data
data(msleep, package = "ggplot2")
library(ggplot2)
library(ggradar)
library(scales)
library(dplyr)
```

```
plotdata <- msleep %>%
  filter(name %in% c("Cow", "Dog", "Pig")) %>%
  select(name, sleep_total, sleep_rem,
         sleep_cycle, brainwt, bodywt) %>%
  rename(group = name) %>%
  mutate_at(vars(-group),
            funs(rescale))
plotdata
```

```
## # A tibble: 3 x 6
##    group sleep_total sleep_rem sleep_cycle brainwt bodywt
##    <chr>       <dbl>     <dbl>       <dbl>   <dbl>  <dbl>
## 1 Cow             0         0           1       1      1
## 2 Dog             1         1           0       0      0
## 3 Pig         0.836     0.773         0.5   0.312  0.123
```

```
# generate radar chart
ggradar(plotdata,
        grid.label.size = 4,
        axis.label.size = 4,
        group.point.size = 5,
        group.line.width = 1.5,
        legend.text.size= 10) +
  labs(title = "Mammals, size, and sleep")
```

In the previous code, the `mutate_at` function rescales all variables except *group*. The various `size` options control the font sizes for the percent labels, variable names, point size, line width, and legend labels respectively.

We can see from Figure 10.15 that cows have relatively larger brain and body weights, longer sleep cycles, and little REM or total sleep time. Dogs in comparison, have small body and brain weights, short sleep cycles, and high total and REM sleep times. (The obvious conclusion is that I want to be a dog–but with a bigger brain).

10.7   Scatterplot Matrix

A scatterplot matrix is a collection of scatterplots (Section 5.2.1) organized as a grid. You can create a scatterplot matrix using the `ggpairs` function in the **GGally** package.

Mammals, size, and sleep

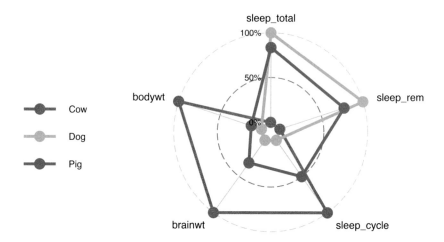

FIGURE 10.15
Basic radar chart.

Let's illustrate its use by examining the relationships between mammal size and sleep characteristics using `msleep` dataset. Brain weight and body weight are highly skewed (think mouse and elephant) so we'll transform them to log brain weight and log body weight before creating the graph.

```
library(GGally)

# prepare data
data(msleep, package="ggplot2")
library(dplyr)
df <- msleep %>%
  mutate(log_brainwt = log(brainwt),
         log_bodywt = log(bodywt)) %>%
  select(log_brainwt, log_bodywt, sleep_total, sleep_rem)


# create a scatterplot matrix
ggpairs(df)
```

By default,

- the principal diagonal contains the kernel density charts (Section 4.2.2) for each variable.

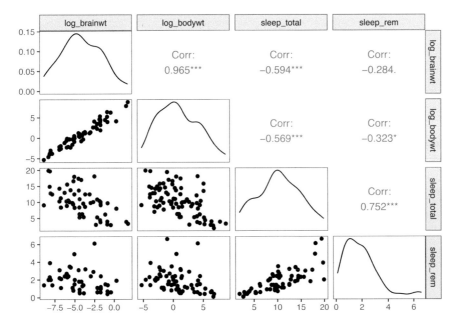

FIGURE 10.16
Scatterplot matrix.

- The cells below the principal diagonal contain the scatterplots represented by the intersection of the row and column variables. The variables across the top are the *x*-axis and the variables down the right side are the *y*-axis.
- The cells above the principal diagonal contain the correlation coefficients.

From Figure 10.16, you can see that as brain weight increases, total sleep time and time in REM sleep decrease. REM sleep times are right skewed (more observations on the lower end, with a few higher scores).

The graph can be modified by creating custom functions.

```
# custom function for density plot
my_density <- function(data, mapping, ...){
  ggplot(data = data, mapping = mapping) +
    geom_density(alpha = 0.5,
                 fill = "cornflowerblue", ...)
}

# custom function for scatterplot
my_scatter <- function(data, mapping, ...){
  ggplot(data = data, mapping = mapping) +
    geom_point(alpha = 0.5,
```

```
                    color = "cornflowerblue") +
     geom_smooth(method=lm,
                     se=FALSE, ...)
}


# create scatterplot matrix
ggpairs(df,
         lower=list(continuous = my_scatter),
         diag = list(continuous = my_density)) +
   labs(title = "Mammal size and sleep characteristics") +
   theme_bw() +
   theme(panel.grid.major=element_blank(),
         panel.grid.minor=element_blank())
```

The results are displayed in Figure 10.17. Being able to write your own functions provides a great deal of flexibility. Additionally, since the resulting plot is a **ggplot2** graph, addition functions can be added to alter the theme, title, labels, etc. See the `?ggpairs` for more details.

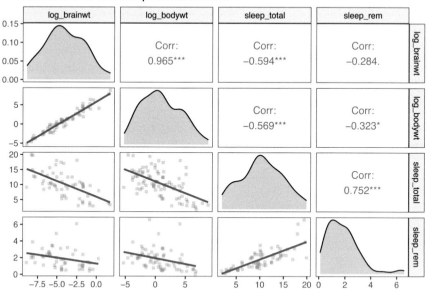

FIGURE 10.17
Customized scatterplot matrix.

10.8   Waterfall Charts

A waterfall chart illustrates the cumulative effect of a sequence of positive and negative values.

For example, we can plot the cumulative effect of revenue and expenses for a fictional company. First, let's create a dataset

```
# create company income statement
category <- c("Sales", "Services", "Fixed Costs",
              "Variable Costs", "Taxes")
amount <- c(101000, 52000, -23000, -15000, -10000)
income <- data.frame(category, amount)
```

Now we can visualize this with a waterfall chart, using the `waterfall` function in the **waterfalls** package.

```
# create waterfall chart
library(ggplot2)
library(waterfalls)
waterfall(income)
```

The result is a basic (default) waterfall chart (Figure 10.18). We can add a total (net) column if desired. Additionally, since the result is a **ggplot2** graph, we can use additional functions to customize the results.

```
# create waterfall chart with total column
waterfall(income,
          calc_total=TRUE,
          total_axis_text = "Net",
          total_rect_text_color="black",
          total_rect_color="goldenrod1") +
  scale_y_continuous(label=scales::dollar) +
  labs(title = "West Coast Profit and Loss",
       subtitle = "Year 2017",
       y="",
       x="") +
  theme_minimal()
```

The customized graph is given in Figure 10.19). Waterfall charts are particularly useful when you want to show change from a starting point to an end point and when there are positive and negative values.

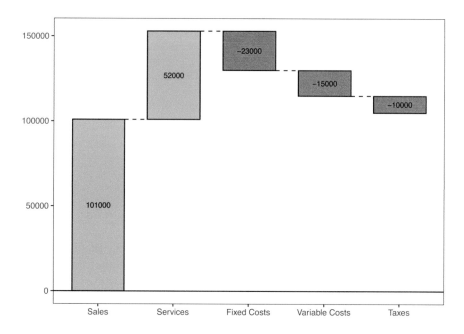

FIGURE 10.18
Basic waterfall chart.

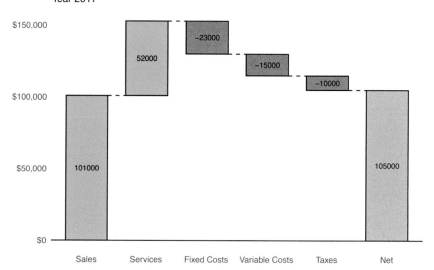

FIGURE 10.19
Waterfall chart with total column.

10.9   Word Clouds

A word cloud (also called a tag cloud), is basically an infographic that indicates the frequency of words in a collection of text (e.g., tweets, a text document, a set of text documents). There is a very nice script produced by STHDA (http://www.sthda.com/english/) that will generate a word cloud directly from a text file.

To demonstrate, we'll use President Kennedy's Address (Appendix A.17) during the Cuban Missile crisis.

To use the script, there are several packages you need to install first. The were not mentioned earlier because they are only needed for this section.

```
# install packages for text mining
install.packages(c("tm", "SnowballC",
                   "wordcloud", "RColorBrewer",
                   "RCurl", "XML"))
```

Once the packages are installed, you can run the script on your text file.

```
# create a word cloud
script <- "http://www.sthda.com/upload/rquery_wordcloud.r"
source(script)
res<-rquery.wordcloud("JFKspeech.txt",
                      type ="file",
                      lang = "english",
                      textStemming=FALSE,
                      min.freq=3,
                      max.words=200)
```

First, the script

- coverts each word to lowercase
- removes numbers, punctuation, and whitespace
- removes stopwords (common words such as "a", "and", and "the")
- if the `textStemming = TRUE` (default is FALSE), words are stemmed (reducing words such as cats, and catty to cat)
- counts the number of times each word appears
- drops words that appear less than three times (*min.freq*)

The script then plots up to 200 words (*max.words*) with word size proportional to the number of times the word appears. The word cloud appears in Figure 10.20.

As you can see, the most common words in the speech are *soviet, cuba, world, weapons,* etc. The terms *missile* and *ballistic* are used rarely.

FIGURE 10.20
Word cloud.

The `rquery.wordcloud` function supports several languages, including Danish, Dutch, English, Finnish, french, German, Italian, Norwegian, Portuguese, Russian, Spanish, and Swedish! See http://www.sthda.com/english/wiki/word-cloud-generator-in-r-one-killer-function-to-do-everything-you-need for details.

11

Customizing Graphs

Graph defaults are fine for quick data exploration, but when you want to publish your results to a blog, paper, article, or poster, you'll probably want to customize the results. Customization can improve the clarity and attractiveness of a graph.

This chapter describes how to customize a graph's axes, gridlines, colors, fonts, labels, and legend. It also describes how to add annotations (text and lines). The last section describes how to combine two of graphs together into one composite image.

11.1 Axes

The x- and y-axis represent numeric, categorical, or date values. You can modify the default scales and labels with the functions below.

11.1.1 Quantitative Axes

A quantitative axis is modified using the `scale_x_continuous` or `scale_y_continuous` function.

Options include

- `breaks`–a numeric vector of positions

- `limits`–a numeric vector with the min and max for the scale

```
# customize numerical x and y axes
library(ggplot2)
ggplot(mpg, aes(x=displ, y=hwy)) +
  geom_point() +
  scale_x_continuous(breaks = seq(1, 7, 1),
                     limits=c(1, 7)) +
  scale_y_continuous(breaks = seq(10, 45, 5),
                     limits=c(10, 45))
```

The resulting plot is given in Figure 11.1.

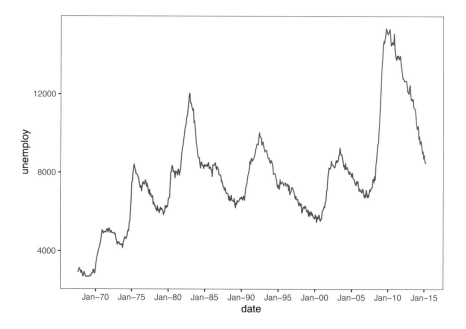

FIGURE 11.4
Customized date axis.

11.2.1   Specifying Colors Manually

To specify a color for points, lines, or text, use the `color = "colorname"`
option in the appropriate geom. To specify a color for bars and areas, use the
`fill = "colorname"` option.
    Examples:

- `geom_point(color = "blue")`

- `geom_bar(fill = "steelblue")`

    Colors can be specified by name or hex code (https://r-charts.com/colors/).
    To assign colors to the levels of a variable, use the `scale_color_manual`
and `scale_fill_manual` functions. The former is used to specify the colors
for points and lines, while the later is used for bars and areas.
    Here is an example, using the `diamonds` dataset that ships with **ggplot2**.
The dataset contains the prices and attributes of 54,000 round cut diamonds.

```
# specify fill color manually
library(ggplot2)
ggplot(diamonds, aes(x = cut, fill = clarity)) +
  geom_bar() +
```

```
scale_fill_manual(values = c("darkred", "steelblue",
                             "darkgreen", "gold",
                             "brown", "purple",
                             "grey", "khaki4"))
```

The plot is given in Figure 11.5. If you are aesthetically challenged like me, an alternative is to use a predefined palette.

11.2.2   Color Palettes

There are *many* predefined color palettes available in R.

11.2.2.1   RColorBrewer

The most popular alternative palettes are probably the ColorBrewer palettes (Figures 11.6 and 11.7).

You can specify these palettes with the `scale_color_brewer` and `scale_fill_brewer` functions.

```
# use an ColorBrewer fill palette
ggplot(diamonds, aes(x = cut, fill = clarity)) +
```

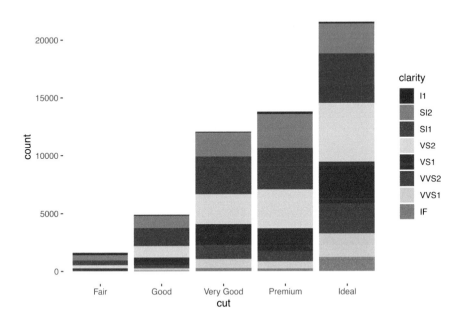

FIGURE 11.5
Stacked barchart (manual color selection).

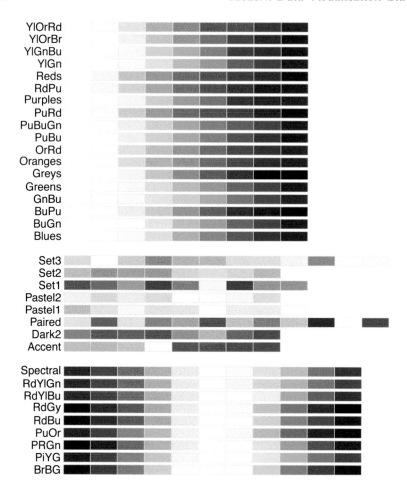

FIGURE 11.6
RColorBrewer palettes.

```
geom_bar() +
scale_fill_brewer(palette = "Dark2")
```

Adding `direction = -1` to these functions reverses the order of the colors in a palette.

11.2.2.2 Viridis

The viridis palette is another popular choice (Figure 11.8). The color scales are attractive, print well in black and white, and more easily perceived by the color blind.

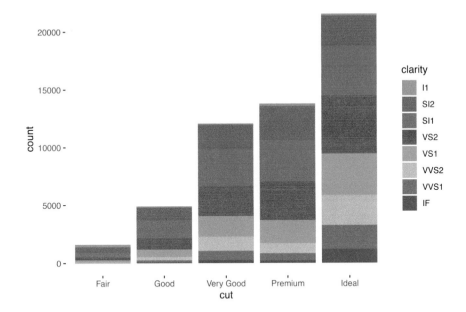

FIGURE 11.7
Stacked barchart (RColorBrewer palette).

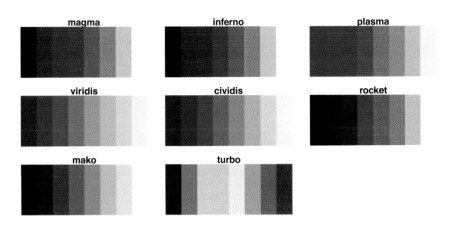

FIGURE 11.8
Viridis palettes.

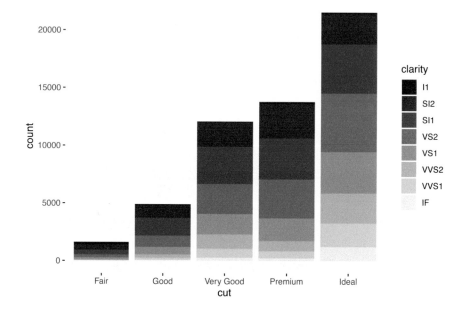

FIGURE 11.9
Stacked barchart (Viridis palette).

For continuous scales use

- `scale_fill_viridis_c`

- `scale_color_viridis_c`

  For discrete (categorical scales) use

- `scale_fill_viridis_d`

- `scale_color_viridis_d`

Here is the previous graph (Figure 11.7), plotted with the viridis palette (Figure 11.9).

```
# Use a viridis fill palette
ggplot(diamonds, aes(x = cut, fill = clarity)) +
  geom_bar() +
  scale_fill_viridis_d()
```

11.2.2.3   Other Palettes

Other palettes to explore include

Package	URL
dutchmasters	https://github.com/EdwinTh/dutchmasters
ggpomological	https://github.com/gadenbuie/ggpomological
LaCroixColoR	https://github.com/johannesbjork/LaCroixColoR
nord	https://github.com/jkaupp/nord
ochRe	https://github.com/ropenscilabs/ochRe
palettetown	https://github.com/timcdlucas/palettetown
pals	https://github.com/kwstat/pals
rcartocolor	https://github.com/Nowosad/rcartocolor
wesanderson	https://github.com/karthik/wesanderson

If you want to explore **all** the palette options (or nearly all), take a look at the **paletter** (https://github.com/EmilHvitfeldt/paletteer) package.

To learn more about color specifications, see the *R Cookpage* page on ggplot2 colors (http://www.cookbook-r.com/Graphs/Colors_(ggplot2)/). For advice on selecting colors, see Section 14.3.

11.3 Points and Lines

11.3.1 Points

For **ggplot2** graphs, the default point is a filled circle. To specify a different shape, use the `shape = #` option in the `geom_point` function. To map shapes to the levels of a categorical variable use the `shape = variablename` option in the `aes` function.

Examples:

- `geom_point(shape = 1)`
- `geom_point(aes(shape = sex))`

Available shapes are given in the Figure 11.10.
Shapes 21 through 26 provide for both a fill color and a border color.

11.3.2 Lines

The default line type is a solid line. To change the linetype, use the `linetype = #` option in the `geom_line` function. To map linetypes to the levels of a categorical variable use the `linetype = variablename` option in the `aes` function.

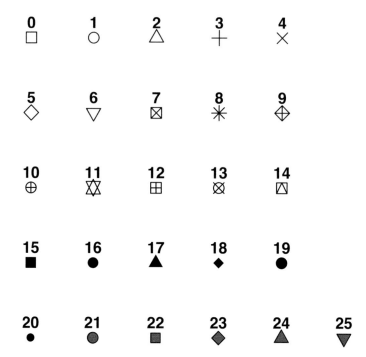

FIGURE 11.10
Point shapes.

Examples:

- `geom_line(linetype = 1)`
- `geom_line(aes(linetype = sex))`

Availabe linetypes are given in Figure 11.11.

11.4   Fonts

R does not have great support for fonts, but with a bit of work, you can change the fonts that appear in your graphs. First you need to install and set-up the `extrafont` package.

```
# one time install
install.packages("extrafont")
library(extrafont)
font_import()
```

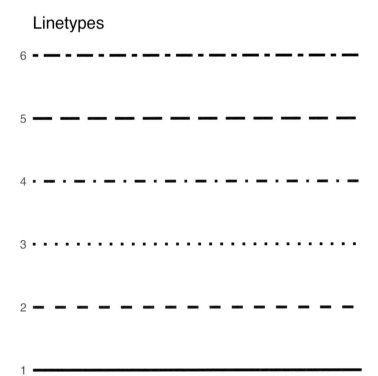

FIGURE 11.11
Linetypes.

```
# see what fonts are now available
fonts()
```

Apply the new font(s) using the `text` option in the `theme` function.

```
# specify new font
library(extrafont)
ggplot(mpg, aes(x = displ, y=hwy)) +
  geom_point() +
  labs(title = "Diplacement by Highway Mileage",
       subtitle = "MPG dataset") +
  theme(text = element_text(size = 16, family = "Comic Sans MS"))
```

To learn more about customizing fonts, see Andrew Heiss's blog on *Working with R, Cairo graphics, custom fonts, and ggplot* (https://www.andrewheiss.com/blog/2017/09/27/working-with-r-cairo-graphics-custom-fonts-and-ggplot/#windows).

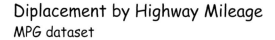

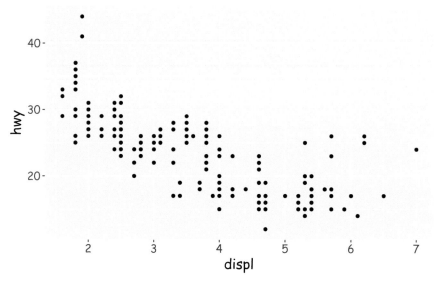

FIGURE 11.12
Alternative fonts.

11.5   Legends

In **ggplot2**, legends are automatically created when variables are mapped to color, fill, linetype, shape, size, or alpha.

You have a great deal of control over the look and feel of these legends. Modifications are usually made through the `theme` function and/or the `labs` function. Here are some of the most sought after changes.

11.5.1   Legend Location

The legend can appear anywhere in the graph. By default, it is placed on the right. You can change the default with

```
theme(legend.position = position)
```
where

Position	Location
"top"	Above the plot area
"right"	Right of the plot area
"bottom"	Below the plot area

"left"	Left of the plot area
c(x, y)	Within the plot area. The x and y values must range between 0 and 1. c(0,0) represents (left, bottom) and c(1,1) represents (right, top).
"none"	Suppress the legend

For example, to place the legend at the top, use the following code. The results are shown in Figure 11.13.

```
# place legend on top
ggplot(mpg,
       aes(x = displ, y=hwy, color = class)) +
  geom_point(size = 4) +
  labs(title = "Diplacement by Highway Mileage") +
  theme_minimal() +
  theme(legend.position = "top")
```

11.5.2 Legend Title

You can change the legend title through the `labs` function. Use `color`, `fill`, `size`, `shape`, `linetype`, and `alpha` to give new titles to the corresponding legends.

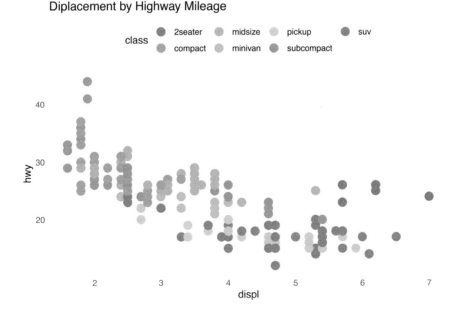

FIGURE 11.13
Moving the legend to the top.

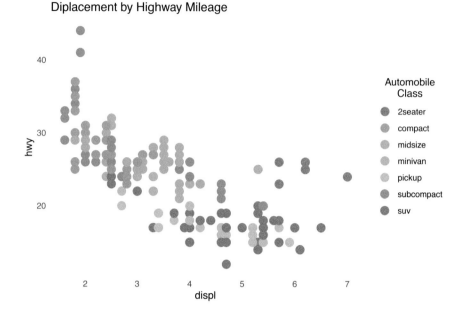

FIGURE 11.14
Changing the legend title.

The alignment of the legend title is controlled through the `legend.title.align` option in the `theme` function (0=left, 0.5=center, 1=right). The code below creates Figure 11.14.

```
# change the default legend title
ggplot(mpg,
       aes(x = displ, y=hwy, color = class)) +
  geom_point(size = 4) +
  labs(title = "Diplacement by Highway Mileage",
       color = "Automobile\nClass") +
  theme_minimal() +
  theme(legend.title.align=0.5)
```

11.6 Labels

Labels are a key ingredient in rendering a graph understandable. They are added with the `labs` function. Available options are given below.

Option	Use
title	Main title
subtitle	Subtitle
caption	Caption (bottom right by default)
x	Horizontal axis
y	Vertical axis
color	Color legend title
fill	Fill legend title
size	Size legend title
linetype	Linetype legend title
shape	Shape legend title
alpha	Transparency legend title
size	Size legend title

For example, consider the following code and graph it creates (Figure 11.15).

```
# add plot labels
ggplot(mpg,
       aes(x = displ, y=hwy,
           color = class,
           shape = factor(year))) +
  geom_point(size = 3,
             alpha = .5) +
  labs(title = "Mileage by engine displacement",
       subtitle = "Data from 1999 and 2008",
       caption = "Source: EPA (http://fueleconomy.gov)",
       x = "Engine displacement (litres)",
       y = "Highway miles per gallon",
       color = "Car Class",
       shape = "Year") +
  theme_minimal()
```

This is not a great graph–it is too busy, making the identification of patterns difficult. It would better to facet the year variable, the class variable or both (Section 6.2). Trend lines would also be helpful (Section 5.2.1.1).

11.7   Annotations

Annotations are additional information added to a graph to highlight important points.

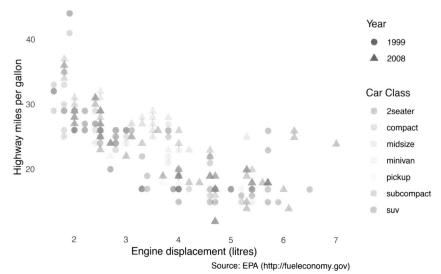

FIGURE 11.15
Graph with labels.

11.7.1   Adding Text

There are two primary reasons to add text to a graph.

One is to identify the numeric qualities of a geom. For example, we may want to identify points with labels in a scatterplot, or label the heights of bars in a bar chart.

Another reason is to provide additional information. We may want to add notes about the data, point out outliers, etc.

11.7.1.1   Labeling Values

Consider the following scatterplot, based on the car data in the mtcars dataset (Figure 11.16).

```
# basic scatterplot
data(mtcars)
ggplot(mtcars, aes(x = wt, y = mpg)) +
  geom_point()
```

Let's label each point with the name of the car it represents.

```
# scatterplot with labels
data(mtcars)
```

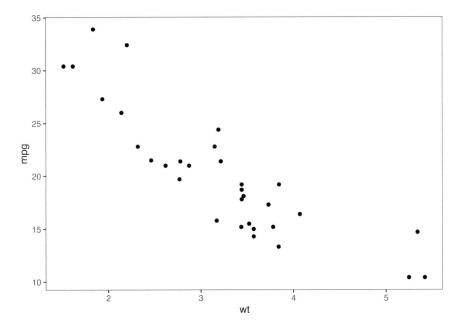

FIGURE 11.16
Simple scatterplot.

```
ggplot(mtcars, aes(x = wt, y = mpg)) +
  geom_point() +
  geom_text(label = row.names(mtcars))
```

The overlapping labels in Figure 11.17 make this chart difficult to read. The ggrepel package can help us here. It nudges text to avoid overlaps.

```
# scatterplot with non-overlapping labels
data(mtcars)
library(ggrepel)
ggplot(mtcars, aes(x = wt, y = mpg)) +
  geom_point() +
  geom_text_repel(label = row.names(mtcars),
                  size=3)
```

Figure 11.18 is much better.

Adding labels to bar charts is covered in the aptly named *labeling bars* section (Section 4.1.1.3).

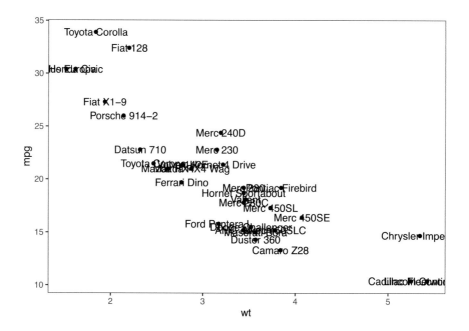

FIGURE 11.17
Scatterplot with labels.

11.7.1.2   Adding Additional Information

We can place text anywhere on a graph using the `annotate` function. The format is

```
annotate("text",
         x, y,
         label = "Some text",
         color = "colorname",
         size=textsize)
```

where x and y are the coordinates on which to place the text. The `color` and `size` parameters are optional.

By default, the text will be centered. Use `hjust` and `vjust` to change the alignment:

- `hjust` 0 = left justified, 0.5 = centered, and 1 = right centered.
- `vjust` 0 = above, 0.5 = centered, and 1 = below.

Continuing the previous example, let's modify the point colors and add explanatory text (Figure 11.19).

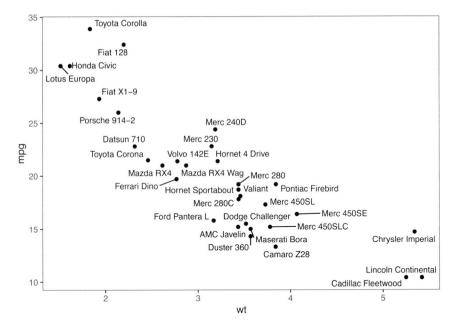

FIGURE 11.18
Scatterplot with non-overlapping labels.

```
# scatterplot with explanatory text
data(mtcars)
library(ggrepel)
txt <- paste("The relationship between car weight",
             "and mileage appears to be roughly linear",
             sep = "\n")
ggplot(mtcars, aes(x = wt, y = mpg)) +
  geom_point(color = "red") +
  geom_text_repel(label = row.names(mtcars),
                  size=3) +
  ggplot2::annotate("text",
                    6, 30,
                    label=txt,
                    color = "red",
                    hjust = 1) +
  theme_bw()
```

   See this stackoverflow blog post (https://stackoverflow.com/questions/
7263849/what-do-hjust-and-vjust-do-when-making-a-plot-using-ggplot) for
more details.

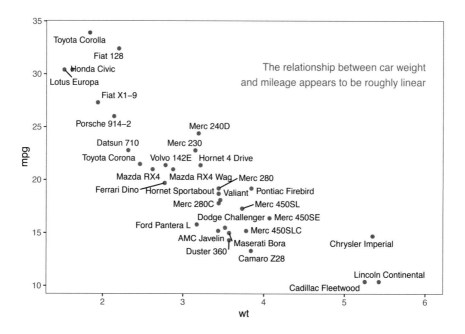

FIGURE 11.19
Annotated scatterplot with adjusted labels.

11.7.2   Adding Lines

Horizontal and vertical lines can be added using:

- `geom_hline(yintercept = a)`
- `geom_vline(xintercept = b)`

where a is a number on the y-axis and b is a number on the x-axis, respectively. Other options include `linetype` and `color`.

In the following example, we plot city vs. highway miles and indicate the mean highway miles with a horizontal line and label.

```
# add annotation line and text label
min_cty <- min(mpg$cty)
mean_hwy <- mean(mpg$hwy)
ggplot(mpg,
       aes(x = cty, y=hwy, color=drv)) +
  geom_point(size = 3) +
  geom_hline(yintercept = mean_hwy,
             color = "darkred",
             linetype = "dashed") +
  ggplot2::annotate("text",
```

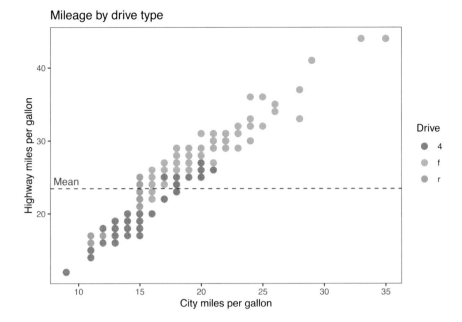

FIGURE 11.20
Graph with line annotation.

```
            min_cty,
            mean_hwy + 1,
            label = "Mean",
            color = "darkred") +
   labs(title = "Mileage by drive type",
        x = "City miles per gallon",
        y = "Highway miles per gallon",
        color = "Drive")
```

The results are displayed in Figure 11.20. We could add a vertical line for the mean city miles per gallon as well. In any case, always label your annotation lines in some way. Otherwise the reader will not know what they mean.

11.7.3   Highlighting a Single Group

Sometimes you want to highlight a single group in your graph. The `gghighlight` function in the `gghighlight` package is designed for this.

Here is an example with a scatterplot (Figure 11.21). Midsize cars are highlighted.

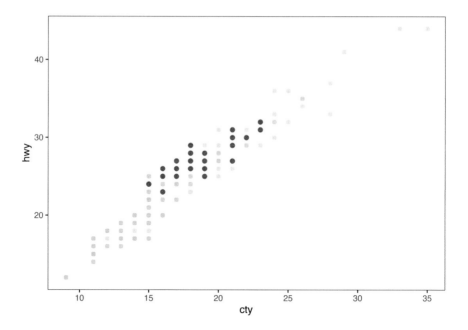

FIGURE 11.21
Highlighting a group.

```
# highlight a set of points
library(ggplot2)
library(gghighlight)
ggplot(mpg, aes(x = cty, y = hwy)) +
  geom_point(color = "red",
             size=2) +
  gghighlight(class == "midsize")
```

Below is an example with a bar chart (Figure 11.22). Again, midsize cars are highlighted.

```
# highlight a single bar
library(gghighlight)
ggplot(mpg, aes(x = class)) +
  geom_bar(fill = "red") +
  gghighlight(class == "midsize")
```

Highlighting is helpful for drawing the reader's attention to a particular group of observations and their standing with respect to the other observations in the data.

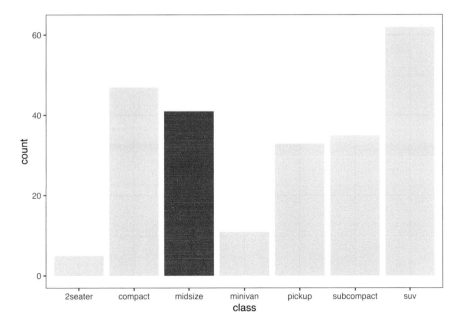

FIGURE 11.22
Highlighting a group.

11.8 Themes

ggplot2 themes control the appearance of all non-data related components of a plot. You can change the look and feel of a graph by altering the elements of its theme.

11.8.1 Altering Theme Elements

The **theme** function is used to modify individual components of a theme.

Consider the following code and graph (Figure 11.23). It shows the number of male and female faculty by rank and discipline at a particular university in 2008–2009. The data come from the **salaries** dataset in the **carData** package.

```
# create graph
data(Salaries, package = "carData")
p <- ggplot(Salaries, aes(x = rank, fill = sex)) +
  geom_bar() +
  facet_wrap(~discipline) +
  labs(title = "Academic Rank by Gender and Discipline",
```

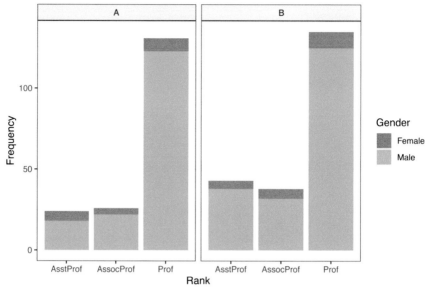

FIGURE 11.23
Graph with default theme.

```
        x = "Rank",
        y = "Frequency",
        fill = "Gender")
p
```

Let's make some changes to the theme.

- Change label text from black to navy blue
- Change the panel background color from grey to white
- Add solid grey lines for major *y*-axis grid lines
- Add dashed grey lines for minor *x*-axis grid lines
- Eliminate *x*-axis grid lines
- Change the strip background color to white with a grey border

Using the ?theme help in **ggplot2** gives us

```
p +
  theme(text = element_text(color = "navy"),
      panel.background = element_rect(fill = "white"),
      panel.grid.major.y = element_line(color = "grey"),
      panel.grid.minor.y = element_line(color = "grey",
```

```
                                            linetype = "dashed"),
  panel.grid.major.x = element_blank(),
  panel.grid.minor.x = element_blank(),
  strip.background = element_rect(fill = "white",
                                   color="grey"))
```

The resulting plot is given in Figure 11.24. Wow, this looks pretty awful, but you get the idea.

11.8.1.1   ggplot Theme Assistant

If you would like to create your own theme using a GUI, take a look at the **ggThemeAssist** package. After you install the package, a new menu item will appear under Addins in RStudio (see Figure 11.25).

Highlight the code that creates your graph, then choose the **ggplot Theme Assistant** option from the **Addins** drop-down menu. You can change many of the features of your theme using point-and-click. When you're done, the **theme** code will be appended to your graph code.

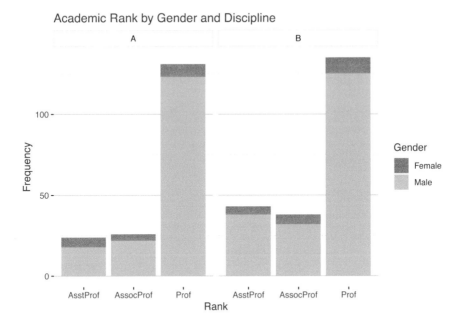

FIGURE 11.24
Graph with modified theme.

FIGURE 11.25
ggplot Theme Assistant.

11.8.2   Pre-Packaged Themes

I'm not a very good artist (just look at the previous example), so I often look for pre-packaged themes that can be applied to my graphs. There are many available.

Some come with **ggplot2**. These include *theme_classic, theme_dark, theme_gray, theme_grey, theme_light theme_linedraw, theme_minimal,* and *theme_void.* We've used *theme_minimal* often in this book. Others are available through add-on packages.

11.8.2.1   ggthemes

The `ggthemes` package come with 19 themes.

Theme	Description
theme_base	Theme Base
theme_calc	Theme Calc
theme_economist	ggplot color theme based on the Economist
theme_economist_white	ggplot color theme based on the Economist
theme_excel	ggplot color theme based on old Excel plots
theme_few	Theme based on Few's "Practical Rules for Using Color in Charts"
theme_fivethirtyeight	Theme inspired by fivethirtyeight.com plots
theme_foundation	Foundation Theme
theme_gdocs	Theme with Google Docs Chart defaults
theme_hc	Highcharts JS theme

Theme	Description
theme_igray	Inverse gray theme
theme_map	Clean theme for maps
theme_pander	A ggplot theme originated from the pander package
theme_par	Theme which takes its values from the current "base" graphics parameter values in "par"
theme_solarized	ggplot color themes based on the Solarized palette
theme_solarized_2	ggplot color themes based on the Solarized palette
theme_solid	Theme with nothing other than a background color
theme_stata	Themes based on Stata graph schemes
theme_tufte	Tufte Maximal Data, Minimal Ink Theme
theme_wsj	Wall Street Journal theme

To demonstrate their use, we'll first create and save a graph (Figure 11.26).

```
# create basic plot
library(ggplot2)
p <- ggplot(mpg,
            aes(x = displ, y=hwy,
                color = class)) +
  geom_point(size = 3,
             alpha = .5) +
  labs(title = "Mileage by engine displacement",
       subtitle = "Data from 1999 and 2008",
       caption = "Source: EPA (http://fueleconomy.gov)",
       x = "Engine displacement (litres)",
       y = "Highway miles per gallon",
       color = "Car Class")

# display graph
p
```

Now, let's apply some themes. We will try the Economist theme (Figure 11.27), the Five Thirty Eight theme (Figure 11.28), the Wall Street Journal theme (Figure 11.29), and the Few theme (Figure 11.30).

```
# add economist theme
library(ggthemes)
p + theme_economist()


# add fivethirtyeight theme
p + theme_fivethirtyeight()
```

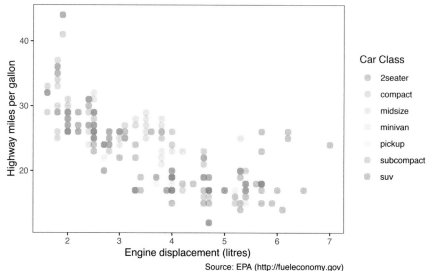

FIGURE 11.26
Default theme.

```
# add wsj theme
p + theme_wsj(base_size=8)
```

By default, the font size for the wsj theme is usually too large. Changing the `base_size` option can help.

Each theme also comes with scales for colors and fills. In the next example, both the `few` theme and colors are used.

```
# add few theme
p + theme_few() + scale_color_few()
```

Try out different themes and scales to find one that you like.

11.8.2.2   hrbrthemes

The `hrbrthemes` package is focused on typography-centric themes. The results are charts that tend to have a clean look.

Continuing the example plot from above

```
# add few theme
library(hrbrthemes)
p + theme_ipsum()
```

The result is given in Figure 11.29. See the hrbrthemes homepage (https://github.com/hrbrmstr/hrbrthemes) for additional examples.

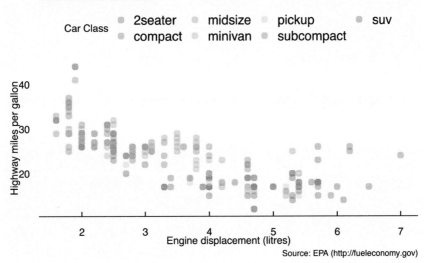

FIGURE 11.27
Economist theme.

11.8.2.3 ggthemer

The `ggthemer` package offers a wide range of themes (17 as of this printing).
The package is not available on CRAN and must be installed from GitHub.

```
# one time install
install.packages("remotes")
remotes::install_github('cttobin/ggthemr')
```

The functions work a bit differently. Use the `ggthemr("themename")` function to set future graphs to a given theme. Use `ggthemr_reset()` to return future graphs to the **ggplot2** default theme.

Current themes include *flat, flat dark, camoflauge, chalk, copper, dust, earth, fresh, grape, grass, greyscale, light, lilac, pale, sea, sky,* and *solarized.* Let's try the flat dark theme (shown in Figure 11.30).

```
# set graphs to the flat dark theme
library(ggthemr)
ggthemr("flat dark")
p
```

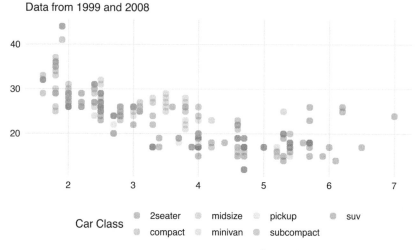

FIGURE 11.28
Five Thirty Eight theme.

```
ggthemr_reset()
```

I wouldn't actually use this theme for this particular graph. It is difficult to distinguish colors. Which green represents compact cars and which represents subcompact cars?

Select a theme that best conveys the graph's information to your audience.

11.9   Combining Graphs

At times, you may want to combine several graphs together into a single image. Doing so can help you describe several relationships at once. The **patchwork** package can be used to combine ggplot2 graphs into a mosaic and save the results as a **ggplot2** graph.

First save each graph as a **ggplot2** object. Then combine them using | to combine graphs horizontally and / to combine graphs vertically. You can use parentheses to group graphs.

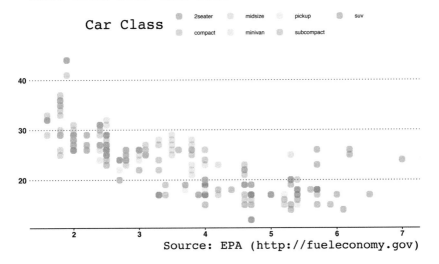

FIGURE 11.29
Wall Street Journal theme.

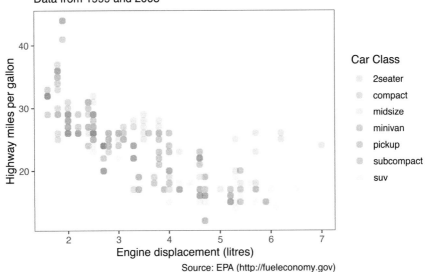

FIGURE 11.30
Few theme and colors.

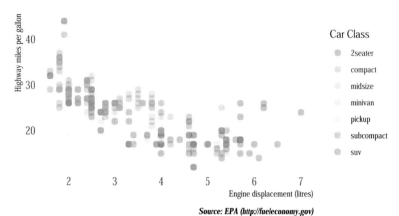

FIGURE 11.31
Ipsum theme.

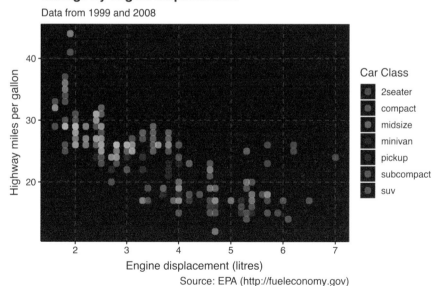

FIGURE 11.32
Flat Dark theme.

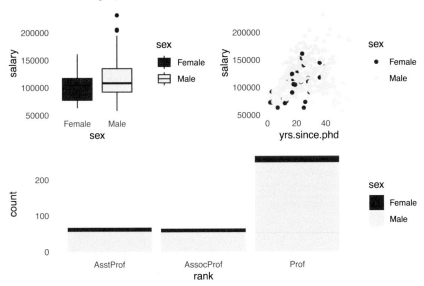

FIGURE 11.33
Combining graphs using the patchwork package.

Here is an example using the `Salaries` dataset from the **carData** package. The combined plot will display the relationship between sex, salary, experience, and rank.

```
data(Salaries, package = "carData")
library(ggplot2)
library(patchwork)

# boxplot of salary by sex
p1 <- ggplot(Salaries, aes(x = sex, y = salary, fill=sex)) +
  geom_boxplot()

# scatterplot of salary by experience and sex
p2 <- ggplot(Salaries,
             aes(x = yrs.since.phd, y = salary, color=sex)) +
  geom_point()

# barchart of rank and sex
p3 <- ggplot(Salaries, aes(x = rank, fill = sex)) +
    geom_bar()
```

```
# combine the graphs and tweak the theme and colors
(p1 | p2)/p3 +
  plot_annotation(title = "Salaries for college professors") &
  theme_minimal() &
  scale_fill_viridis_d() &
  scale_color_viridis_d()
```

The `plot_annotation` function allows you to add a title and subtitle to
the entire graph. Note that the `&` operator applies a function to *all* graphs in
a plot. If we had used `+ theme_minimal()` only the bar chart (the last graph)
would have been affected. The results are given in Figure 11.31.

The patchwork package allows for exact placement and sizing of graphs,
and even supports insets (placing one graph within another). See https://
patchwork.data-imaginist.com for details.

12

Saving Graphs

Graphs can be saved via the RStudio interface or through code.

12.1 Via Menus

To save a graph using the RStudio menus, go to the **Plots** tab and choose Export (Figure 12.1).

12.2 Via Code

Any **ggplot2** graph can be saved as an object. Then you can use the `ggsave` function to save the graph to disk.

```
# save a graph
library(ggplot2)
p <- ggplot(mtcars,
            aes(x = wt , y = mpg)) +
  geom_point()
ggsave(p, filename = "mygraph.png")
```

The graph will be saved in the format defined by the file extension (*png* in the example above). Common formats are *pdf, jpeg, tiff, png, svg,* and *wmf* (windows only).

DOI: 10.1201/9781003299271-12

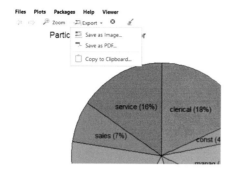

FIGURE 12.1
RStudio image export menu.

12.3 File Formats

Graphs can be saved in several formats. The most popular choices are given below.

Format	Extension
Portable document format	pdf
JPEG	jpeg
Tagged image file format	tiff
Portable network graphics	png
Scaleable vector graphics	svg
Windows metafile	wmf

The *pdf*, *svg*, and *wmf* formats are lossless–they resize without fuzziness or pixelation. The other formats are lossy–they will pixelate when resized. This is especially noticeable when small images are enlarged.

If you are creating graphs for webpages, the *png* format is recommended. The *jpeg* and *tif* formats are usually reserved for photographs.

The *wmf* format is usually recommended for graphs that will appear in Microsoft Word or PowerPoint documents. MS Office does not support *pdf* or *svg* files, and the *wmf* format will rescale well. However, note that *wmf* files will lose any transparency settings that have been set.

If you want to continue editing the graph after saving it, use the *pdf* or *svg* format.

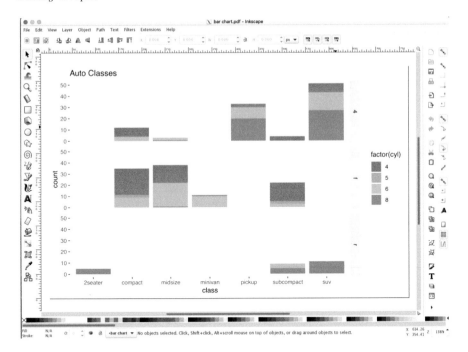

FIGURE 12.2
Inkscape.

12.4 External Editing

Sometimes it's difficult to get a graph just right programmatically. Most magazines and newspapers (print and electronic) fine-tune graphs after they have been created. They change the fonts, move labels around, add callouts, change colors, add additional images or logos, and the likes.

If you save the graph in *svg* or *pdf* format, you can use a vector graphics editing program to modify it using point and click tools. Two popular vector graphics editors are **Illustrator** and **Inkscape**.

Inkscape (https://inkscape.org) is an opensource application that can be freely downloaded for Mac OS X, Windows, and Linux. A screenshot of the interface is provided in Figure 12.2. Open the graph file in *Inkscape*, edit it to suite your needs, and save it in the format desired.

13

Interactive Graphs

Interactive graphs allow for greater exploration and reader engagement. With the exception of maps (Section 7) and 3-D scatterplots (Section 10.1), this book has focused on static graphs–images that can be placed in papers, posters, slides, and journal articles. Through connections with JavaScript libraries, such as **htmlwidgets for R** (https://www.htmlwidgets.org), R can generate interactive graphs that can be explored in RStudio's viewer window or placed on external web pages.

This chapter will explore several approaches including **plotly**, **ggiraph**, **rbokeh**, **rcharts**, and **highcharter**. The focus is on simple, straight-forward approaches to adding interactivity to graphs. Be sure to run the code so that you can experience the interactivity. Alternatively, the online version of this book (http://rkabacoff.github.io/datavis) offers an interactive version of each graph.

The **Shiny** framework offers a comprehensive approach to interactivity in R (https://www.rstudio.com/products/shiny/). However, **Shiny** has a higher learning curve and requires access to a **Shiny** server, so it is not considered here. Interested readers are referred to this excellent text (Sievert, 2020).

13.1 plotly

Plotly (https://plot.ly/) is both a commercial service and open source product for creating high end interactive visualizations. The **plotly** package allows you to create plotly interactive graphs from within R. In addition, any **ggplot2** graph can be turned into a plotly graph.

Using the `mpg` data that comes with the **ggplot2** package, we'll create an interactive graph displaying highway mileage vs. engine displace by car class. The **ggplot2** package is used to create and save the graph as an object. The `ggplotly` function in the **plotly** package takes this object and renders it as interactive graph (Figure 13.1).

DOI: 10.1201/9781003299271-13

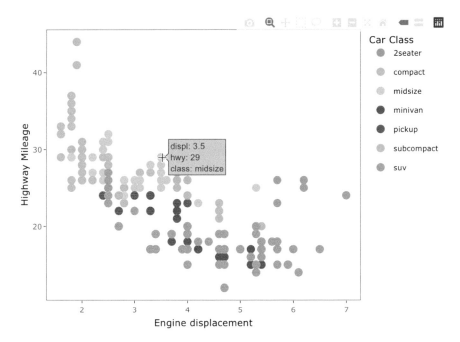

FIGURE 13.1
Plotly graph.

```
# create plotly graph.
library(ggplot2)
library(plotly)

p <- ggplot(mpg, aes(x=displ,
                     y=hwy,
                     color=class)) +
  geom_point(size=3) +
  labs(x = "Engine displacement",
       y = "Highway Mileage",
       color = "Car Class") +
  theme_bw()

ggplotly(p)
```

Mousing over a point displays information about that point. Clicking on a legend point, removes that class from the plot. Clicking on it again returns it. Popup tools on the upper right of the plot allow you to zoom in and out of the image, pan, select, reset axes, and download the image as a *png* file.

By default, the mouse over provides pop-up tooltip with values used to create the plot (*dipl*, *hwy*, and *class* here). However, you can customize the tooltip. This involves adding a label$n = variablen$* to the `aes` function and to the `ggplotly` function.

```r
# create plotly graph.
library(ggplot2)
library(plotly)

p <- ggplot(mpg, aes(x=displ,
                     y=hwy,
                     color=class,
                     label1 = manufacturer,
                     label2 = model,
                     label3 = year)) +
  geom_point(size=3) +
  labs(x = "Engine displacement",
       y = "Highway Mileage",
       color = "Car Class") +
  theme_bw()

ggplotly(p, tooltip = c("label1", "label2", "label3"))
```

The tooltip now displays the car manufacturer, make, and year (see Figure 13.2).

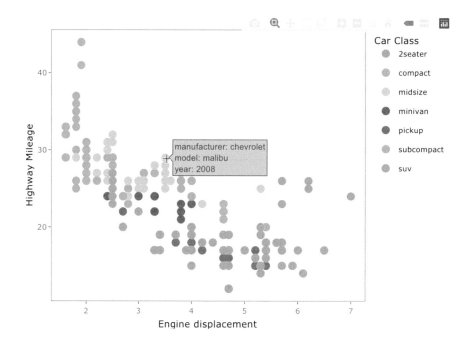

FIGURE 13.2
Plotly graph with custom tooltip.

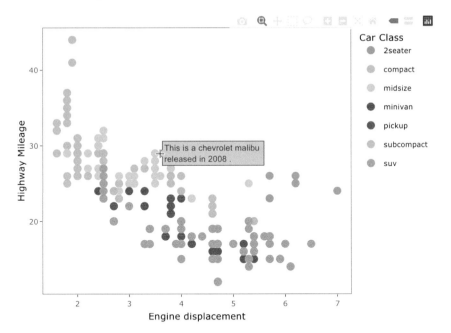

FIGURE 13.3
Plotly graph with fully customized tooltip.

You can fully customize the tooltip by creating your own label and including it as a variable in the data frame. Then place it in the aesthetic as text and in the ggplotly function as a label. The code is given below and the graph appears in Figure 13.3.

```
# create plotly graph.
library(ggplot2)
library(plotly)
library(dplyr)

mpg <- mpg %>%
  mutate(mylabel = paste("This is a", manufacturer, model, "\n",
                         "released in", year, "."))

p <- ggplot(mpg, aes(x=displ,
                     y=hwy,
                     color=class,
                     text = mylabel)) +
  geom_point(size=3) +
  labs(x = "Engine displacement",
       y = "Highway Mileage",
```

```
            color = "Car Class") +
  theme_bw()

ggplotly(p, tooltip = c("mylabel"))
```

There are several sources of good information on plotly. See the *plotly R pages* (https://plot.ly/r/) and the book *Interactive web-based data visualization with R, Plotly, and Shiny* (Sievert, 2020). An online version of the book is available at https://plotly-book.cpsievert.me/.

13.2   ggiraph

It is easy to create interactive **ggplot2** graphs using the `ggiraph` package. There are three steps:

1. add `_interactive` to the geom names (for example, change *geom_point* to *geom_point_interactive*)
2. add `tooltip`, `data_id`, or both to the `aes` function
3. render the plot with the `girafe` function (note the spelling)

First, let's use **ggiraph** to create an interactive scatter plot between engine displacement and highway mileage.

```
library(ggplot2)
library(ggiraph)

p <- ggplot(mpg, aes(x=displ,
                y=hwy,
                color=class,
                tooltip = manufacturer)) +
  geom_point_interactive()

girafe(ggobj = p)
```

In Figure 13.4 when you mouse over a point, the manufacturer's name pops up. The ggiraph package only allows one tools tip, but you can customize it by creating a column in the data containing the desired information. By default, the tooltip is white text on a black background. You can change this with the `options` argument to the `girafe` function.

In the next example, a tooltip is created by concatenating the *manufacturer*, *model*, and *class* variables using the `paste` function, and saving the results in the dataset as the variable *tooltip*. In the `girafe` function, the code

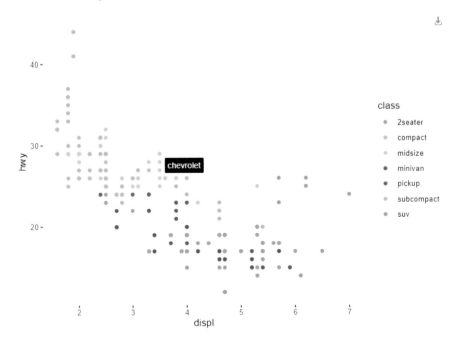

FIGURE 13.4
Basic interactive ggiraph graph.

opts_tooltip(use_fill = TRUE) sets the tooltip to the fill color (*class* in this case). The resulting graph is displayed in Figure 13.5.

```
library(ggplot2)
library(ggiraph)
library(patchwork)

library(dplyr)
mpg <- mpg %>%
  mutate(tooltip = paste(manufacturer, model, class))

p <- ggplot(mpg, aes(x=displ,
                y=hwy,
                color=class,
                tooltip = tooltip)) +
  geom_point_interactive()

girafe(ggobj = p, options = list(opts_tooltip(use_fill = TRUE)))
```

Section 11.9 described how to combine two or more **ggplot2** graphs into one over all plot using the **patchwork** package. One of the great strengths of

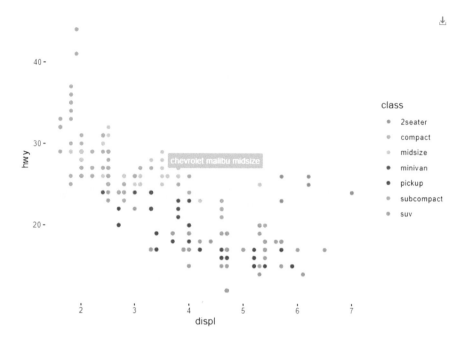

FIGURE 13.5
Interactive ggiraph graph with customized tooltip.

the **ggiraph** package is that it allows you to **link** these graphs. When graphs
are linked, selecting observations on one graph highlights the same observations
in each of the other graphs.

The next example uses the fuel efficiency data from the `mtcars` dataset.
Three plots are created (1) a scatterplot of weight vs mpg, (2) a scatterplot of
rear axle ratio vs 1/4 mile time, and (3) a bar chart of number of cylinders.

Unlike previous graphs, these three graphs are linked by a unique id
(the car names in this case). This is accomplished by adding `data_id =
rownames(mtcars)` to the aesthetic. The three plots are then arranged into
one graph using *patchwork* and `code=print` is added to the *girafe* function.

In Figure 13.6, clicking on location in any one of the three plots, highlights
the car in each of the other plots. Run the code and try it out!

```
library(patchwork)
p1 <- ggplot(mtcars, aes(x=wt,
                    y=mpg,
                    tooltip = rownames(mtcars),
                    data_id = rownames(mtcars))) +
   geom_point_interactive(size=3, alpha =.6)
```

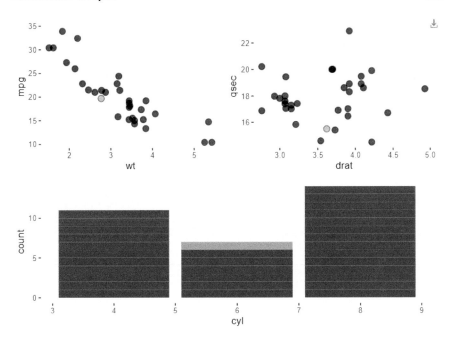

FIGURE 13.6
Linked interactive graphs.

```
p2 <- ggplot(mtcars, aes(x=drat,
                         y=qsec,
                         tooltip = rownames(mtcars),
                         data_id = rownames(mtcars))) +
   geom_point_interactive(size = 3, alpha = .6)

p3 <- ggplot(mtcars, aes(x=cyl,
                         data_id = rownames(mtcars))) +
   geom_bar_interactive()

p3 <- (p1 | p2)/p3
girafe(code = print (p3))
```

Here is one more example. The `gapminder` dataset is used to create two bar charts for Asian countries: (1) life expectancy in 1982 and life expectancy in 2007. The plots are linked by `data_id = country`. As you mouse over a bar in one chart, the corresponding bar in the other is highlighted (see Figure 13.7). If you move from the bottom to the top in the left hand chart, it become clear how life expectancy has changed. Note the jump when you hit Vietnam and Iraq.

Life Expectancy in Asia

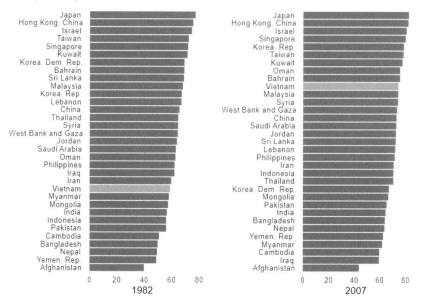

FIGURE 13.7
Linked bar charts.

```
data(gapminder, package="gapminder")

# subset Asian countries
asia <- gapminder %>%
  filter(continent == "Asia") %>%
  select(year, country, lifeExp)

p1 <- ggplot(asia[asia$year == 1982,],
       aes(y = reorder(country, lifeExp),
           x=lifeExp,
           tooltip = lifeExp,
           data_id = country)) +
  geom_bar_interactive(stat="identity",
                       fill="steelblue") +
  labs(y="", x="1982") +
  theme_minimal()

p2 <- ggplot(asia[asia$year == 2007,],
             aes(y = reorder(country, lifeExp),
                 x=lifeExp,
```

```
                    tooltip = lifeExp,
                    data_id = country)) +
  geom_bar_interactive(stat="identity",
                       fill="steelblue") +
  labs(y="", x="2007") +
  theme_minimal()

p3 <- (p1 | p2) +
  plot_annotation(title = "Life Expectancy in Asia")
girafe(code = print (p3))
```

Graphs created with ggiraph are highly customizable. The ggiraph-book website (https://www.ardata.fr/ggiraph-book/) is a great resource for getting started.

13.3   Other Approaches

While **Plotly** is the most popular approach for turning static **ggplot2** graphs into interactive plots, many other approaches exist. Describing each in detail is beyond the scope of this book. Examples of other approaches are included here in order to give you a taste of what each is like. You can then follow the references to learn more about the ones that interest you.

13.3.1   rbokeh

rbokeh is an interface to the **Bokeh** (https://bokeh.pydata.org/en/latest/) graphics library.

We'll create another graph using the mtcars dataset, showing engine displace vs. miles per gallon by number of engine cylinders. Mouse over Figure 13.8 and try the various control to the right of the image.

```
# create rbokeh graph

# prepare data
data(mtcars)
mtcars$name <- row.names(mtcars)
mtcars$cyl <- factor(mtcars$cyl)

# graph it
library(rbokeh)
figure() %>%
  ly_points(disp, mpg, data=mtcars,
```

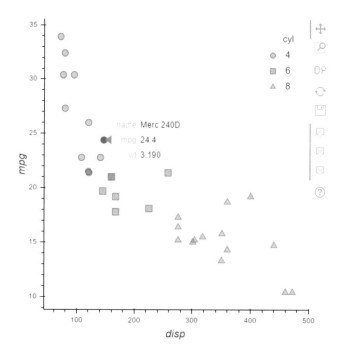

FIGURE 13.8
Bokeh graph.

```
color = cyl, glyph = cyl,
hover = list(name, mpg, wt))
```

You can create some remarkable graphs with Bokeh. See the homepage
(http://hafen.github.io/rbokeh/) for examples.

13.3.2   rCharts

rCharts can create a wide range of interactive graphics. In the example below,
a bar chart of hair vs. eye color is created (Figure 13.9). Try mousing over
the bars. You can interactively choose between grouped vs. stacked plots and
include or exclude cases by eye color by clicking on the legends at the top of
the image.

```
# create interactive bar chart
library(rCharts)
hair_eye_male = subset(as.data.frame(HairEyeColor),
                       Sex == "Male")
n1 <- nPlot(Freq ~ Hair,
```

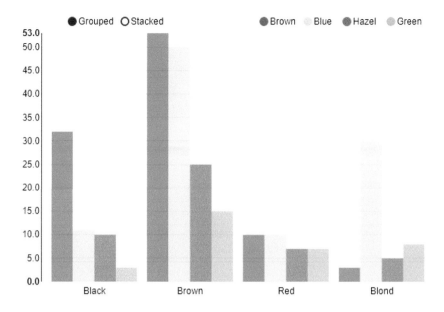

FIGURE 13.9
rCharts graph.

```
                group = 'Eye',
                data = hair_eye_male,
                type = 'multiBarChart'
)
n1$set(width = 600)
n1$show('iframesrc', cdn=TRUE)
```

To learn more, visit the project homepage (https://github.com/ramnathv/rCharts).

13.3.3 highcharter

The **highcharter** package provides access to the *Highcharts* (https://www.highcharts.com/) JavaScript graphics library. The library is free for noncommercial use.

Let's use **highcharter** to create an interactive line chart displaying life expectancy over time for several Asian countries. The data come from the `gapminder` dataset. Again, mouse over the lines and try clicking on the legend names.

```
# create interactive line chart
library(highcharter)

# prepare data
data(gapminder, package = "gapminder")
library(dplyr)
asia <- gapminder %>%
  filter(continent == "Asia") %>%
  select(year, country, lifeExp)

# convert from long to wide format
library(tidyr)
plotdata <- pivot_wider(asia,
                        names_from = "country",
                        values_from = "lifeExp")

# generate graph
h <- highchart() %>%
  hc_xAxis(categories = plotdata$year) %>%
  hc_add_series(name = "Afghanistan",
                data = plotdata$Afghanistan) %>%
  hc_add_series(name = "Bahrain",
                data = plotdata$Bahrain) %>%
  hc_add_series(name = "Cambodia",
                data = plotdata$Cambodia) %>%
  hc_add_series(name = "China",
                data = plotdata$China) %>%
  hc_add_series(name = "India",
                data = plotdata$India) %>%
  hc_add_series(name = "Iran",
                data = plotdata$Iran)

h
```

In Figure 13.10, I've clicked on the Afghanistan point in 1962. The line is highlighted, the other lines are dimmed, and a pop-up box shows the values at that point.

Like all of the interactive graphs in this chapter, there are options that allow the graph to be customized. The code below adds styled titles, a caption, and a legend (see Figure 13.11). The tooltip background color is set to a pale yellow. And when mousing over a point, *all* points in that year are highlighted and a pop-up menu shows each country's values for that year.

```
# customize interactive line chart
h <- h %>%
```

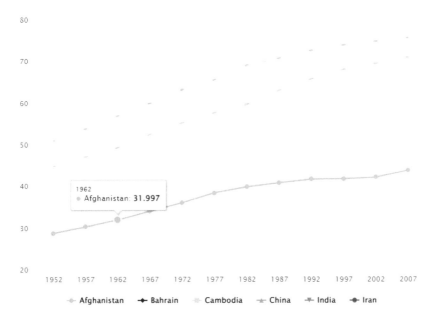

FIGURE 13.10
HighCharts graph.

```
hc_title(text = "Life Expectancy by Country",
         margin = 20,
         align = "left",
         style = list(color = "steelblue")) %>%
hc_subtitle(text = "1952 to 2007",
            align = "left",
            style = list(color = "#2b908f",
                         fontWeight = "bold")) %>%
hc_credits(enabled = TRUE, # add credits
           text = "Gapminder Data",
           href = "http://gapminder.com") %>%
hc_legend(align = "left",
          verticalAlign = "top",
          layout = "vertical",
          x = 0,
          y = 100) %>%
hc_tooltip(crosshairs = TRUE,
           backgroundColor = "#FCFFC5",
           shared = TRUE,
           borderWidth = 4) %>%
```

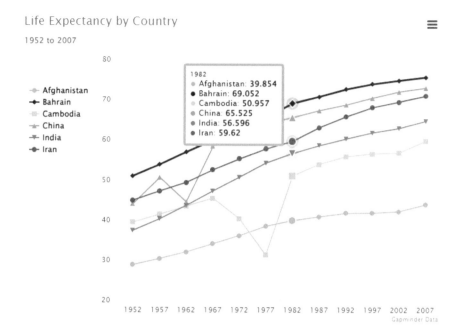

FIGURE 13.11
HighCharts graph with customization.

```
hc_exporting(enabled = TRUE)
```

h

There is a wealth of interactive plots available through the marriage of R and JavaScript. Choose the approach that works best for you.

14

Advice Best Practices

This section contains some thoughts on what makes a good data visualization. Most come from books and posts that others have written, but I'll take responsibility for putting them here.

14.1 Labeling

Everything on your graph should be clearly labeled. Typically this will include a

- *title*–a clear short title letting the reader know what they're looking at
    - *Relationship between experience and wages by gender*

- *subtitle*–an optional second (smaller font) title giving additional information
    - *Years 2016–2018*
- *caption*–source attribution for the data
    - *source: US Department of Labor–www.bls.gov/bls/blswage.htm*
- *axis labels*–clear labels for the x and y axes
    - short but descriptive
    - include units of measurement
        * *Engine displacement (cu. in.)*
        * *Survival time (days)*
        * *Patient age (years)*
- *legend*–short informative title and labels
    - *Male* and *Female*–not 0 and 1 !!
- *lines* and *bars*–label any trend lines, annotation lines, and error bars

Basically, the reader should be able to understand your graph without having to wade through paragraphs of text. When in doubt, show your data visualization to someone who has not read your article or poster and ask them if anything is unclear.

DOI: 10.1201/9781003299271-14

14.2 Signal-to-Noise-Ratio

In data science, the goal of data visualization is to communicate information. Anything that doesn't support this goal should be reduced or eliminated.

> **Chart Junk**–visual elements of charts that aren't necessary to comprehend the information represented by the chart or that distract from this information. (Wikipedia (https://en.wikipedia.org/wiki/Chartjunk))

Consider the graph in Figure 14.1. The goal is to compare the calories in bacon to the other four foods. The data are available at http://blog.cheapism.com. I increased the serving size for bacon from 1 slice to 3 slices (let's be real, its BACON!).

> Disclaimer: I got the idea for this graph from one I saw on the internet years ago, but I can't remember where. If you know, let me know so that I can give proper credit.

If the goal is to compare the calories in bacon to other breakfast foods, much of this visualization is unnecessary and distracts from the task.

Think of all the things that are superfluous:

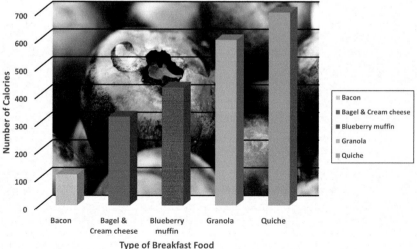

FIGURE 14.1
Graph with chart junk.

- the speckled blue background border
- the blueberries photo image
- the 3-D effect on the bars
- the legend (it doesn't add anything, the bars are already labeled)
- the colors of bars (they don't signify anything)

An alternative plot is given in Figure 14.2.
The chart junk has been removed. In addition,

- the x-axis label isn't needed–these are obviously foods
- the y-axis is given a better label
- the title has been simplified (the word *different* is redundant)
- the bacon bar is the only colored bar—it makes comparisons easier
- the grid lines have been made lighter (gray rather than black) so they don't distract
- calorie values have been added to each bar so that the reader doesn't have to keep refering to the y-axis.

I may have gone a bit far leaving out the x-axis label. It's a fine line, knowing when to stop simplifying.

In general, you want to reduce chart junk to a minimum. In other words, **more signal, less noise**.

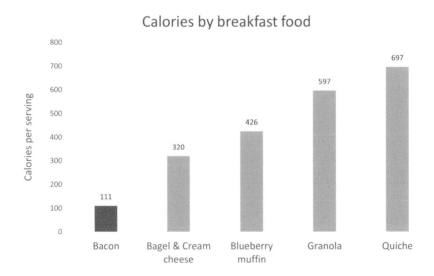

FIGURE 14.2
Graph with chart junk removed.

14.3   Color Choice

Color choice is about more than aesthetics. Choose colors that help convey the information contained in the plot.

The article *How to Choose Colors for Data Visualizations* by Mike Yi (https://chartio.com/learn/charts/how-to-choose-colors-data-visualization) is a great place to start.

Basically, think about selecting among sequential, diverging, and qualitative color schemes:

- Sequential–for plotting a quantitative variable that goes from low to high,
- Diverging–for contrasting the extremes (low, medium, and high) of a quantitative variable.
- Qualitative–for distinguishing among the levels of a categorical variable.

The article above can help you to choose among these schemes. Additionally, the `RColorBrewer` package (Section 11.2.2.1) provides palettes categorized in this way. The *YlOrRd* to *Blues* palettes are sequential, *Set3* to *Accent* are qualitative, and *Spectral* to *BrBg* are diverging.

Other things to keep in mind:

- Make sure that text is legible–avoid dark text on dark backgrounds, light text on light backgrounds, and colors that clash in a discordant fashion (i.e., they hurt to look at!).
- Avoid combinations of red and green–it can be difficult for a colorblind audience to distinguish these colors.

Other helpful resources are Stephen Few's *Practical Rules for Using Color in Charts* (http://www.perceptualedge.com/articles/visual_business_intelligence/rules_for_using_color.pdf) and Maureen Stone's *Expert Color Choices for Presenting Data* (https://courses.washington.edu/info424/2007/documents/Stone-Color%20Choices.pdf).

14.4   *y*-Axis Scaling

OK, this is a big one. You can make an effect seem massive or insignificant depending on how you scale a numeric *y*-axis.

Consider following example, comparing the 9-month salaries of male and female assistant professors. The data come from the `Salaries` dataset.

```r
# load data
data(Salaries, package="carData")

# get means, standard deviations, and
# 95% confidence intervals for
# assistant professor salary by sex
library(dplyr)
df <- Salaries %>%
  filter(rank == "AsstProf") %>%
  group_by(sex) %>%
  summarize(n = n(),
            mean = mean(salary),
            sd = sd(salary),
            se = sd / sqrt(n),
            ci = qt(0.975, df = n - 1) * se)

df

## # A tibble: 2 x 6
##   sex        n   mean    sd    se    ci
##   <fct>  <int>  <dbl> <dbl> <dbl> <dbl>
## 1 Female    11 78050. 9372. 2826. 6296.
## 2 Male      56 81311. 7901. 1056. 2116.

# create and save the plot
library(ggplot2)
p <- ggplot(df,
            aes(x = sex, y = mean, group=1)) +
  geom_point(size = 4) +
  geom_line() +
  scale_y_continuous(limits = c(77000, 82000),
                     label = scales::dollar) +
  labs(title = "Mean salary differences by gender",
       subtitle = "9-mo academic salary in 2007-2008",
       caption = paste("source: Fox J. and Weisberg, S. (2011)",
                       "An R Companion to Applied Regression,",
                       "Second Edition Sage"),
       x = "Gender",
       y = "Salary") +
  scale_y_continuous(labels = scales::dollar)
```

First, let's plot this with a *y*-axis going from 77,000 to 82,000 (Figure 14.3).

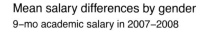

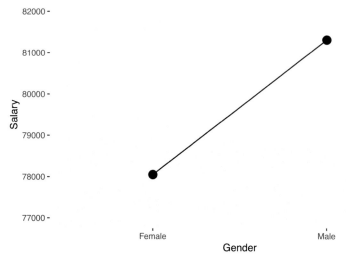

source: Fox J. and Weisberg, S. (2011) An R Companion to Applied Regression, Second Edition Sage

FIGURE 14.3
Plot with limited range of Y.

```
# plot in a narrow range of y
p + scale_y_continuous(limits=c(77000, 82000))
```

There appears to be a very large gender difference.

Next, let's plot the same data with the *y*-axis going from 0 to 125,000. The results are given in Figure 14.4.

```
# plot in a wide range of y
p + scale_y_continuous(limits = c(0, 125000))
```

There doesn't appear to be any gender difference!

The goal of ethical data visualization is to represent findings with as little distortion as possible. This means choosing an appropriate range for the *y*-axis. Bar charts should almost always start at $y = 0$. For other charts, the limits really depends on a subject matter knowledge of the expected range of values.

We can also improve the graph by adding in an indicator of the uncertainty (see the section on Mean/SE plots).

```
# plot with confidence limits
p +  geom_errorbar(aes(ymin = mean - ci,
                       ymax = mean + ci),
                   width = .1) +
```

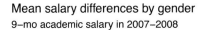

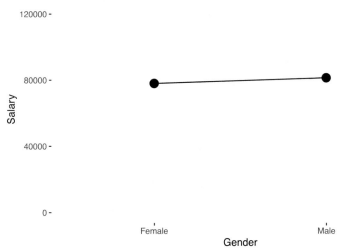

FIGURE 14.4
Plot with limited range of Y.

```
ggplot2::annotate("text",
          label = "I-bars are 95% \nconfidence intervals",
          x=2,
          y=73500,
          fontface = "italic",
          size = 3)
```

As you can see from Figure 14.5, the difference doesn't appear to exceeds chance variation.

14.5 Attribution

Unless it's your data, each graphic should come with an attribution–a note directing the reader to the source of the data. This will usually appear in the caption for the graph.

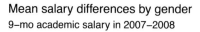

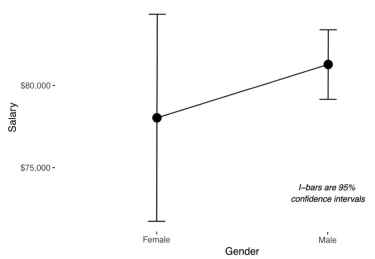

source: Fox J. and Weisberg, S. (2011) An R Companion to Applied Regression, Second Edition Sage

FIGURE 14.5
Plot with error bars.

14.6   Going Further

If you would like to learn more about **ggplot2** there are several good sources, including

- the **ggplot2** homepage (https://ggplot2.tidyverse.org)
- *ggplot2: Elegant graphics for data anaysis (2nd ed.)* (Wickham, 2016). A draft of the third edition is available at https://ggplot2-book.org.
- Chapter 3 in *R for data science* (Wickham & Grolemund, 2017). An online version is available at https://r4ds.had.co.nz/data-visualisation.html.
- the **ggplot2** cheatsheet (https://posit.co/resources/cheatsheets/)

If you would like to learn more about data visualization, in general, here are some useful resources:

- Scott Berinato's Harvard Business Review article *Visualizations that really work* (https://hbr.org/2016/06/visualizations-that-really-work)
- *Wall Street Journal's guide to information graphics: The dos and don'ts of presenting data, facts and figures* (Wong, 2010)
- *A practical guide to graphics reporting : Information graphics for print, web & broadcast* (George-Palilonis, 2017)

- *Beautiful data: The stories behind elegant data solutions* (Segaran & Hammerbacher, 2009)
- *The truthful art: Data, charts, and maps for communication* (Cairo, 2016)
- The *Information is beautiful* website (https://informationisbeautiful.net)

The best graphs are rarely created on the first attempt. Experiment until you have a visualization that clarifies the data and helps communicates a meaning story. And have fun!

A

Datasets

The appendix describes the datasets used in this book.

A.1   Academic Salaries

The `Salaries` dataset comes from the `carData` package. It describes the 9-month academic salaries of 397 college professors at a single institution in 2008–2009. The data were collected as part of the administration's monitoring of gender differences in salary.

The dataset can be accessed using

```
data(Salaries, package="carData")
```

The dataset is provided in several formats on the book support website (http://www.github.com/rkabacoff/datavis), so that you can practice importing these file types.

Format	File
Comma delimited text	Salaries.csv
Tab delimited text	Salaries.txt
Excel spreadsheet	Salaries.xlsx
SAS file	Salaries.sas7bdat
Stata file	Salaries.dta
SPSS file	Salaries.sav

A.2   Star Wars

The `Star Wars` dataset comes from the **dplyr** package. It describes 13 characteristics of 87 characters from the Star Wars universe. The data are extracted from the *Star Wars API* (http://swapi.co).

A.3   Mammal Sleep

The `msleep` dataset comes from the **ggplot2** package. It is an updated and expanded version of a dataset by Save and West, describing the sleeping characteristics of 83 mammals.

The dataset can be accessed using

```
data(msleep, package="ggplot2")
```

A.4   Medical Insurance Costs

The `insurance` dataset is described in the book **Machine Learning with R** by Brett Lantz is is available from the author's website (https://github.com/dataspelunking/MLwR). A version of the dataset is also available on *Kaggle* (https://www.kaggle.com/datasets/mirichoi0218/insurance).

The dataset describes medical information and costs billed by health insurance companies in 2013, as compiled by the United States Census Bureau. Variables include age, sex, body mass index, number of children covered by health insurance, smoker status, US region, and individual medical costs billed by health insurance for 1338 individuals.

A.5   Marriage Records

The `Marriage` dataset comes from the **mosiacData** package. It is contains the marriage records of 98 individuals collected from a probate court in Mobile County, Alabama.

The dataset can be accessed using

```
data(Marriage, package="mosaicData")
```

A.6   Fuel Economy Data

The `mpg` dataset from the **ggplot2** package, contains fuel economy data for 38 popular models of car, for the years 1999 and 2008.

The dataset can be accessed using

```
data(mpg, package="ggplot2")
```

A.7 Literacy Rates

This dataset provides the literacy rates (percent of the population that can both read and write) for each US state in 2023. The data were obtained from the World Population Review (https://worldpopulationreview.com/state-rankings/us-literacy-rates-by-state).

The dataset can be obtained from http://github.com/kabacoff/datavis and imported using

```
library(readr)
litRates <- read_csv("USLitRates.csv")
```

A.8 Gapminder Data

The `gapminder` dataset from the **gapminder** package, contains longitudinal data (1952–2007) on life expectancy, GDP per capita, and population for 142 countries.

The dataset can be accessed using

```
data(gapminder, package="gapminder")
```

A.9 Current Population Survey (1985)

The `CPS85` dataset from the **mosaicData** package, contains 1985 data on wages and other characteristics of workers.

The dataset can be accessed using

```
data(CPS85, package="mosaicData")
```

A.10 Houston Crime Data

The `crime` dataset from the **ggmap** package, contains the time, date, and location of six types of crimes in Houston, Texas between January 2010 and August 2010.

The dataset can be accessed using

```
data(crime, package="ggmap")
```

A.11 Hispanic and Latino Populations

The Hispanic and Latino Populations data is a raw tab delimited text file containing the percentage of Hispanic and Latinos by US state from the 2010 Census. The actual dataset was obtained from Wikipedia (https://en.wikipedia.org/wiki/List_of_U.S._states_by_Hispanic_and_Latino_population).

The dataset can be obtained from http://github.com/kabacoff/datavis and imported using

```
library(readr)
text <- read_csv("hisplat.csv")
```

A.12 US Economic Timeseries

The `economics` dataset from the **ggplot2** package, contains the monthly economic data gathered from Jan 1967 to Jan 2015.

The dataset can be accessed using

```
data(economics, package="ggplot2")
```

A.13 US Population by Age and Year

The uspopage dataset describes the age distribution of the US population from 1900 to 2002.

The dataset can be accessed using

```
data(uspopage, package="gcookbook")
```

A.14 Saratoga Housing Data

The `SaratogaHouses` dataset contains information on 1728 houses in Saratoga Country, NY, USA in 2006. Variables include price (in thousands of US dollars) and 15 property characteristics (lotsize, living area, age, number of bathrooms, etc.)

The dataset can be accessed using

```
data(SaratogaHouses, package="mosaicData")
```

A.15 NCCTG Lung Cancer Data

The `lung` dataset describes the survival time of 228 patients with advanced lung cancer from the North Central Cancer Treatment Group.

The dataset can be accessed using

```
data(lung, package="survival")
```

A.16 Titanic Data

The Titanic dataset provides information on the fate of Titanic passengers, based on class, sex, and age. The dataset comes in table form with base R.

The dataset can be obtained from http://github.com/kabacoff/datavis and imported using

```
library(readr)
titanic <- read_csv("titanic.csv")
```

A.17 JFK Cuban Missle Speech

The John F. Kennedy Address is a raw text file containing the president's October 22, 1962, speech on the Cuban Missle Crisis. The text was obtained from the *JFK Presidential Library and Museum* (https://www.jfklibrary.org/JFK/Historic-Speeches.aspx).

The dataset can be obtained from http://github.com/kabacoff/datavis and imported using

```
library(readr)
text <- read_csv("JFKspeech.txt")
```

B

About the Author

Robert Kabacoff is a data scientist with more than 30 years of experience in multivariate statistical methods, data visualization, predictive analytics, and statistical programming. As a professor of the Practice in Quantitative Analysis at the Hazel Quantitative Analysis Center at Wesleyan University, he teaches courses in applied data analysis, machine learning, data journalism, and advance R programming. Rob is the author of *R in Action: Data Analysis and Graphics with R* (3rd ed.), and a popular website on R programming called Quick-R.

DOI: 10.1201/9781003299271-B

C

About the QAC

The Quantitative Analysis Center (QAC) is a collaborative effort of academic and administrative departments at Wesleyan University. It coordinates support for quantitative analysis across the curriculum, and provides education and research support for both students and faculty.

DOI: 10.1201/9781003299271-C

Bibliography

Breheny, P., & Burchett, W. (2017). Visualization of regression models using visreg. *The R Journal, 9*(2), 56–71. https://doi.org/10.32614/RJ-2017-046

Byron, L., & Wattenberg, M. (2008). Stacked graphs–geometry and aesthetics. *IEEE Transactions on Visualization and Computer Graphics, 14*(6), 1245–1252. https://doi.org/10.1109/TVCG.2008.166

Cairo, A. (2016). *The truthful art : Data, charts, and maps for communication* [Includes bibliographical references and index.]. New Riders.

Cleveland, W. S. (1979). Robust locally weighted regression and smoothing scatterplots. *Journal of the American Statistical Association, 74*(368), 829–836. https://doi.org/10.1080/01621459.1979.10481038

George-Palilonis, J. (2017). *A practical guide to graphics reporting : Information graphics for print, web & broadcast* (2nd ed.) [Includes bibliographical references and index.]. Routledge. https://doi.org/10.4324/9781315709574

Kahle, D., & Wickham, H. (2013). Ggmap: Spatial visualization with ggplot2. *The R Journal, 5*(1), 144–161. https://doi.org/10.32614/rj-2013-014

Kowarik, A., & Templ, M. (2016). Imputation with the R package VIM. *Journal of Statistical Software, 74*(7), 1–16. https://doi.org/10.18637/jss.v074.i07

Lovelace, R., Nowasad, J., & Meuchow, J. (2019). *Geocomputation with r*. Chapman & Hall/CRC.

McGill, R., Tukey, J. W., & Larsen, W. A. (1978). Variations of box plots. *The American Statistician, 32*(1), 12–16. https://doi.org/10.2307/2683468

Nishisato, S., Beh, E. J., Lombardo, R., & Clavel, J. G. (2021). History of the biplot. In S. Nishisato, E. J. Beh, R. Lombardo, & J. G. Clavel (Eds.), *Modern quantification theory: Joint graphical display, biplots, and alternatives* (pp. 167–179). Springer Singapore. https://doi.org/10.1007/978-981-16-2470-4_9

R Core Team. (2023). *R: A language and environment for statistical computing*. R Foundation for Statistical Computing. Vienna, Austria. https://www.R-project.org/

Segaran, T., & Hammerbacher, J. (2009). Beautiful data: The stories behind elegant data solutions.

Sievert, C. (2020). *Interactive web-based data visualization with r, plotly, and shiny*. CRC Press.

Wickham, H. (2016). *Ggplot2 : Elegant graphics for data analysis* (2nd ed.) [Includes bibliographical references at the end of each chapters and index]. Springer International Publishingr.

Wickham, H., Chang, W., Henry, L., Pedersen, T. L., Takahashi, K., Wilke, C., Woo, K., Yutani, H., & Dunnington, D. (2023). *Ggplot2: Create elegant data visualisations using the grammar of graphics* [R package version 3.4.2]. https://CRAN.R-project.org/package=ggplot2

Wickham, H., & Grolemund, G. (2017). *R for data science : Import, tidy, transform, visualize, and model data* [Includes bibliographical references and index.]. O'Reilly.

Wilkinson, L., & Wills, G. (2005). *The grammar of graphics* (2nd ed.) [Includes bibliographical references (pp. 635–671) and indexes]. Springer.

Wong, D. M. (2010). *The wall street journal guide to information graphics: The dos and don'ts of presenting data, facts, and figures* (1st ed.). W.W. Norton & Company.

Index